Practical Watch Collecting

For the Beginner

Richard Watkins

NAWCC

National Association of
WATCH&CLOCK
Collectors, Inc.

Library of Congress Cataloging-in-Publication Data

Watkins, Richard (Richard P.)
    Practical watch collecting for the beginner / Richard Watkins.
        pages cm
    Includes bibliographical references and index.
    ISBN 978-0-9823584-5-0
    1. Clocks and watches--Collectors and collecting. I. Title.
    NK7484.W38 2012
    681.1'14075--dc23
                        2012001089

Printed in the United States of America
The National Association of Watch and Clock Collectors, Inc.
Editor: Diana M. DeLucca; Associate Editors: Freda Conner, Hugh Dougherty, Amy L. Klinedinst

Requests to use material from this work should be directed to:
The National Association of Watch and Clock Collectors, Inc.
514 Poplar Street, Columbia, PA 17512

Founded in 1943, the National Association of Watch and Clock Collectors, Inc. (NAWCC)
is a nonprofit member organization whose purpose is to encourage and stimulate interest
in the art and science of horology for the benefit of NAWCC members and the public.

See the last page of this book for more information about the NAWCC
and a reproducible membership application.

All images in this publication are courtesy of the author unless otherwise noted.

Contents

Preface

An author should have a definite purpose, and his or her writing should reflect this. My purpose is to describe a practical and hopefully enjoyable way for the beginner, with no knowledge, to start on the journey of becoming a serious watch collector. To achieve this I have based this book on my own experiences and how I learned about watches, their history, and mechanisms.

Such a purpose is too vague and needs to be clarified by detailing four specific aims.

First, I decided that the fundamental question to be answered is: What does a beginner want to do? Quite simply, it is to be able to pick up a watch, examine it, and draw conclusions about it, including who made it, how old it is, its quality, and its condition. In practice this is done from the outside in, starting with the external appearance of the case and dial, and progressing inward to the finer details of the movement. This process defines the order in which this book is written.

Of course, a proper and complete assessment of a watch requires considerable experience and knowledge, which the beginner does not have. So the second aim is to limit the book to what is essential for such a person; that which he or she is capable of doing and appreciating. One consequence of this is that I believe learners should only look at and collect ordinary watches. Such people simply do not have the experience and knowledge necessary to evaluate complicated, rare, or expensive watches, and to do so will often lead to grief. So this book concentrates on common watches. Where it looks ahead, I have limited myself to providing just a taste of what is to come.

My third aim is to avoid unnecessary duplication. There are nearly 3,000 books that examine all aspects of watches and watchmaking; many are very good, but others, unfortunately, are poor. Rather than write a voluminous treatise, I have directed the reader to some of these works that, over the years, should become essential parts of a personal library. Because I have no idea what will become the novice collector's interest, it is impossible to create an essential reading guide. I recommend the reader reference my annotated bibliography, *Mechanical Watches*, which contains much information about what is available and, more importantly, its contents and quality.

Finally, it should be noted that a book for beginners has a limited life-span; it is read once or twice and then put aside, as the reader's knowledge and needs develop, and it is replaced by more substantial texts. So such a book should be well produced but inexpensive. (In contrast, "coffee table" books are as opulent and expensive as possible!) Which is why professional photography has not been used and a luxurious (unnecessary) binding has been avoided.

Acknowledgments

I especially thank Michael Edidin for his many suggestions and corrections. I also thank the entire editorial staff of the NAWCC, for their work in converting a rough idea into a book and their patience coping with my seemingly endless requests for changes.

In addition, I thank Andrew Shepherdson, for supplying some of the wristwatches photographed for this book, and Niels Hellgren, for providing the photographs of the MI Chronometer in Chapter 5.

Richard Watkins
Kingston, Tasmania, Australia
www.watkinsr.id.au, books@watkinsr.id.au

Chapter 1

Before We Begin ...

Collectors and Dealers

Before we begin, let me say something about this book; I am doing so here because I suspect many people don't bother to read forewords and prefaces, and I want you to read this.

Some people only collect watches made in the country where they live. Consequently, an American reader, for example, may not be interested in the parts of this book that discuss English watches, and might prefer a book that only examined American watches. However, I know people living in the United States who collect English watches, and they may not be interested in a book that only looks at American watches. I also know people in England who collect American watches. I can't win!

In addition to writing for people living anywhere who collect any type of watch, I believe all collectors should have some knowledge of all sorts of watches, even if they only collect watches from a particular locality or period in time. For example, I am interested in watches made in Liverpool, England, between about 1790 and 1835. But the designs of early American movements were based on these watches, and there are watches with American names and places engraved on them that were made in Liverpool during this period. There are also watches with American names and places engraved on them that were made in Switzerland and other places. So a collector of American watches really needs to know something about English and Swiss watchmaking, as well as about American makers. Indeed, everyone needs at least a basic understanding of watchmaking throughout the world.

Therefore, I request that you do not skip sections that, on the surface, are irrelevant to your main interests. Later, when you have developed enough knowledge and experience, you can specialize to your heart's content.

This book is for collectors like me—people who feel an urgent desire to own something they do not need and to add it to the other things their urgent desire had made them buy previously. Collecting is a passion, ruled by the feelings of pleasure and satisfaction when a new object is found and acquired, interspersed with the occasional periods of depression when the lusted-after object turns out to be less than satisfactory. For we must be realistic and be aware that we all occasionally do something stupid and live to regret it. Like the French watch made circa 1780 with a blue enameled back that I paid far too much for, and I can't bring myself to sell it because I will lose money.

The main problem collectors have is discovering what will become their lifelong passion. As a child I dabbled with model trains, stamps, and books. As an adult, preoccupied with earning a living, I toyed with classical records—Beethoven and the like. But one day I saw a Waltham pocket watch at a flea market, liked it and bought it. I have no idea what attracted me to it, but I have never looked back!

Why people choose to collect one thing rather than another is a mystery—another aspect of personality I suppose. But it most certainly is not rational, let alone sensible. I had an academic education and have never been interested in, let alone good at, things mechanical. So why did I fall in love with what many see as the ultimate expression of mechanical engineering? And love is the right word; collectors do not like the things they collect, they unashamedly love them.

One very important point about the psychology of collecting is what happens when you first discover what is to become your lifelong passion. The very first watch, book, or whatever it is that you choose to collect is endowed with an immense emotional value, no matter how good or important it actually is. My first watch is rather ordinary, but it has a significance in my collection that far outweighs its value. Similarly, the first book you read about watches will seem excellent because it will open your eyes and mind to watches. But just like your first watch, if later you go back to that book, you might find it to be ordinary and uninspiring. Not that your more educated opinion matters, because that book will always hold a special place in your heart.

What Is Watch Collecting?

Watch collecting is the creation of a personal museum that illustrates some aspect of the history of horology.

First, your collection is just the same as one in a public museum, except that you can handle the objects rather than just look at them in a glass display case, and the fact that you can handle them is crucial. The watches you have are a selection that represents what interests you. And just as museum displays change, your collection may change to suit the development of your interests.

Second, your collection will illustrate some aspect of watches. Just as some people only collect Australian postage stamps, after a while you will probably discover that you want to focus on a particular type of watch and not collect a bit of everything.

Third, your collection illustrates the history of horology. Although you may not agree, I think a watch collector must collect watches—the movement, the mechanism, is the focus of watch collecting, not the case, not the bracelet, but the wheels, levers, and other bits that make it tick. Watch collectors do not collect jewelry that happens to have hands. Watch collectors do not collect precious metals shaped into funny little boxes called watchcases. They collect watches. Yes, the watchcase may be made of gold studded with diamonds, but the identical mechanism in a plain steel box would be just as good (almost!). If you want to collect beautiful small boxes, why not snuff boxes, or cigarette cases, or any of a hundred other types of jewelry containers? No, if you want to collect watches then, despite being attracted by a beautiful case, it is the movement you must learn about.

From Novice to Competent Collector

To gain the necessary knowledge and experience to become competent requires four things:

You must have your own library of essential texts.
You must handle as many real watches as possible.
You must be willing to invest a lot of time.
You should belong to one or more associations.

One basic reference book you need to have in your library is my book *Mechanical Watches: An Annotated Bibliography of Publications Since 1800*. If you have access to the Internet, then you can get a free copy by downloading it from *www.watkinsr.id.au*. If you do not have access to the Internet, I hope you know someone who does, or you can persuade a local library to get a copy. I feel a bit embarrassed, recommending my own book, but there is no other bibliography that is useful for beginners and collectors. The important point for you is that many books are reviewed so that you can judge whether they will interest you.

The important books that I recommend you buy for your own library are reviewed in *Mechanical Watches*, and are all books that I consider to be excellent or very good. All the books I mention are in one of the two lists "A Beginner's Library" and "Further Reading" at the end of this book.

Remember that the books I recommend are only a tiny fraction of the more than 2,000 books and pamphlets that have been printed and include information on watches. There are many books that you can choose to read instead of my recommendations. Later you will want to increase your general or specialist knowledge. *Mechanical Watches* will provide you with a guide so that you can choose good books to read rather than taking potluck. Here are a couple of books that you may find useful:

Eric Bruton, *Dictionary of Clocks and Watches*. This is a small book that explains much of the terminology of horology. It is a bit superficial, but it is worth reading once. Try to borrow it rather than buy it. There is a later printing of this book called *Collector's Dictionary of Clocks and Watches,* which is a much expanded edition.

Donald de Carle, *Watch and Clock Encyclopedia*. This is also a dictionary that defines terms, but it is larger and more comprehensive than Bruton's book.

I have not listed these books in my suggestions for a beginner's library because I think they are of limited value. Both define terms, often without explanations, and simply tell you what technical words mean. Although this is important, you will probably find them rather frustrating.

A much more useful book that I do recommend for your personal library is F. J. Britten's *Watch and Clockmakers' Handbook, Dictionary and Guide.* This is a more advanced book than Bruton or de Carle. It is also in dictionary format, but it provides detailed explanations of many things, including escapements. There have been 16 editions of this book, the first in 1878 and the 16th in 1987, and one of the later printings is preferred (I use the 14th or 15th edition, published in 1946 and 1955). It is well worth owning and later I will be referring to it for some specific information.

The problem with dictionaries and encyclopedias is that if you don't know the correct word, you cannot look it up. And if you do know the correct word, you probably don't need to look it up! So, when you first get such a book, you should just sit down and read it from front cover to back to discover the technical words you will need to know.

As your knowledge grows, it will become increasingly valuable for you to belong to an association so that you have access to journals and other collectors. The two English-language associations are:

NAWCC, Inc.
514 Poplar Street
Columbia, PA 17512-2130
USA
http://www.nawcc.org/

The NAWCC produces a journal, the *Watch & Clock Bulletin*, which has been published since 1944, and members can access every issue and a complete, searchable index online. The NAWCC also has an online message board that is an invaluable source of information.

The Antiquarian Horological Society
New House
High Street
Ticehurst
Sussex TN5 7AL
ENGLAND
http://www.ahsoc.demon.co.uk/
The AHS also publishes a journal, *Antiquarian Horology*.

In addition, there are associations based in France and Germany:

Association Française des Amateurs d'Horlogerie Ancienne
http://www.afaha.com/

Deutsche Gesellschaft für Chronometrie
http://www.dg-chrono.de/

Too Late, Too Late!

OK, you are still reading, so I can assume you want to collect watches. But actually I expect that you have already started! Collecting tends to creep up on you and, before you know it, you have several watches. Then you realize that perhaps you ought to know what you are doing.

When I first began to collect watches, every single watch I saw was different and fascinating. I was excited. Because everything was so new, and because I knew absolutely nothing about watches, I bought anything and everything I saw. This is perfectly natural, and I expect everyone goes through a period of excited, unthinking adoration, just like falling in love with another human when everything seems perfect. This period is also a watch dealer's fantasy! I had bought at least 10 watches before it occurred to me that it might be useful to find out something about them.

When people start collecting, they buy what is readily available. For example, Rolex wristwatches are very good, some are not too dear, and there are plenty to choose from. Likewise, there are large numbers of American and English pocket watches on the market. Because there are plenty, the beginner will probably start with cheaper watches and become more selective as his or her knowledge grows. To collect the unusual (e.g., Ormskirk chaffcutters) takes great patience because they are rare—and anyway, the novice probably doesn't even know they exist, let alone what they are. So the process of specializing follows a period of buying virtually anything and developing the experience necessary to realize the areas of horology that interest the individual.

As my collection expanded, I began to realize that I had been a bit stupid and that some of the watches I had bought were best described as junk. Indeed, I think it is inevitable that novice collectors will make some of their worst purchases when they begin. Although this can be embarrassing, it doesn't matter as long as you learn from your mistakes. I also suspect that recurring flashes of lust result in occasional bad buys until you die; you just get used to it being part of the addiction. Later I realized that the "junk" I had bought was a necessary part of the learning process and every watch I had acquired served a useful role in my education.

As my knowledge and experience grew, I started to specialize; no more wristwatches, no more ordinary English pocket watches, but ... But what? I still haven't decided! I like early ninetenth-century

English watches, pocket chronometers, repeaters, and master navigation watches, but I am really not sure what I prefer! Maybe one day I will decide, but for now I still buy what interests me without much of a reason why. There has been one major change in my collecting behavior—I now collect far more books about watches than watches.

By the time you read this, you will almost certainly have started collecting and I have to play catch-up; I have to ask you to stop buying and start reading. Or, if that is too hard, try to do both at once.

At this point one important piece of advice is: When you start collecting watches, it is a good idea to begin with some cheap watches and even, perhaps, some that don't work.

While you are a beginner and your knowledge is limited, it is very risky to buy expensive watches. What you need to do first is to learn enough to be able to make reasonable judgments. A few "junk" watches, given to you or purchased for very little, will give you the opportunity to examine and even pull apart watches without having to worry about damaging them. And a few hours fiddling with movements will give you invaluable experience. Most people like to gamble occasionally. Instead of betting on horses or going to casinos, I gamble on watches. I will happily buy a watch that may not work if it seems interesting and doesn't cost much. Sometimes I find I have wasted my money, but occasionally I get a good watch that just has a few minor problems; it may not be perfect, but it's good enough to add to my collection. Either way, not much money is invested and I attribute the cost to learning. Nearly all such buys have taught me something more about watches. So do not spurn "rubbish," especially in the early stages of your education.

Buying

Because I am playing catch-up, I am going to give some advice on buying watches and books now, even though this topic should come later.

Decide if you want to buy a watch or a book for what you can learn from it or for its appearance and value. When buying watches, the most interesting ones in the best condition are usually the most expensive. Sometimes, like me, you might be happy with a less-than-perfect watch rather than not have one at all. The same applies to books. A first edition in excellent condition might be desirable, but a later edition or a cheaper reprint with a few faults will be just as useful. Indeed, sometimes a later printing has more information and is much better.

Never buy the first watch or book you see. Frequently, there are several similar items for sale at any one time and often at startlingly different prices. After you have spent some time browsing, you will quite often discover that a watch or book that has been touted as rare is, in fact, readily available. So always shop around.

Always form your own opinion of the seller. As much as we would like to imagine that all people are reputable and fair, in reality many are not. Sellers vary from the scrupulously honest to downright liars, and you cannot afford to buy from anyone until you have decided what type of person you are dealing with. This is fairly easy if you meet the person face-to-face and can handle the goods; you can talk and ask questions to assess the person as well as the merchandise. But it requires a good deal of skill when your only contact is by telephone or email.

Never buy a pocket watch unless you have seen the movement or a good photograph of it. Beautiful cases in excellent condition can hide ordinary movements or movements in bad condition. Quite often I have seen interesting watches on the Internet with no photographs of the movement, so I have contacted the seller and asked for one. Always, with no exceptions, when the photograph arrived, I discovered the movement was totally uninteresting, which explains why a picture of it was not supplied in the first place.

I impose the same condition on wristwatches, but often it is impossible to look at the movement without special tools. In addition, opening a waterproof or water-resistant wristwatch case may break the seal and result in it no longer being waterproof, so people are often reluctant to open such cases.

Watch dealers will usually ask the highest prices (in line with published price guides). However, if they are reputable, they should give some kind of guarantee that what they are selling you is "as described" in fine detail. Such a guarantee is essential if you are buying from an Internet watch dealer.

General antique dealers and secondhand shops will usually sell at lower prices. But the seller normally doesn't know much about watches and cannot provide a guarantee. The best you can hope for is that if a watch works and you discover a major fault, you might be able to return it.

Auctions, whether local or on the Internet, are often the cheapest but also the highest risk. What you are willing to bid is entirely up to your knowledge, your evaluation of each watch, and your desire for it.

Internet auction sites, such as eBay, require special care. Many buyers and sellers are ignorant. Ignorant buyers bid erratically, often paying too much on a whim and frustrating knowledgeable bidders. Ignorant sellers provide inadequate information, and sometimes their descriptions are grossly inaccurate. I also suspect some reasonably knowledgeable sellers deliberately suggest they are ignorant so that they can avoid providing disadvantageous descriptions. In contrast, some watch dealers provide detailed information about themselves and almost no information about what they are selling, which always seem to have high prices—well above the normal auction prices of similar items!

I have quite often bought watches from eBay and I have developed a few rules. If the photographs of the item are poor, do not bid. In particular, I will never bid on a pocket watch unless there is at least one very good photograph of the movement. Read the description at least twice. Descriptions are frequently wrong or misleading. Work out what sort of person the seller is. It is usually quite easy to deduce something of the personality of the vendor: ignorant, evasive, honest and open, helpful, trying to overadvertise, and so on. Check other sources. See whether similar items are available elsewhere and what the asking prices are. Decide on a maximum bid and stick to it. At any auction it is easy to be carried away and overbid.

The same points apply to buying horological books.

The best way to buy books is from the Internet. There are several very large Internet sites that sell secondhand and rare books, and you will usually find several copies of a book for sale at any one time. If you are looking for a particular book, use an Internet search site such as www.addall.com/used/. This site automatically searches many large Internet sites that sell secondhand and rare books and returns a list of available copies. If you are not looking for a particular book, but just want to browse through a list to see what exists, read my book *Mechanical Watches: An Annotated Bibliography of Publications Since 1800*. Alternatively, I suggest you try http://www.abebooks.com/servlet/SearchEntry. Search using the key word horology, and sort the results by most recently listed and you will get a random listing of over 6,000 books to browse through!

There are booksellers who specialize in horological books. Four that I know well that focus on English language books are:

Jeffrey Formby Antiques
Orchard Cottage
East Street
Moreton in Marsh
Gloucestershire GL56 0LQ
http://www.formby-clocks.co.uk/

G.K. Hadfield
Old Post Office
Great Salkeld
Penrith
Cumbria CA11 9LN
http://www.gkhadfield-tilly.co.uk/

Rogers Turner Books
87 Breakspears Road
London SE4 1TX
rogersturner@compuserve.com

Shenton Books
2 The Lawns
Cheddar
Somerset BS27 3RD
http://www.shentonbooks.com/

Although there are some small horological booksellers in the United States, and some new booksellers, I don't know of any major ones. There are also horological booksellers in Germany (http://www.uhrenbuch.de/) and Switzerland (http://www.booksimonin.ch/) who focus on books in their respective languages.

Terminology

One challenge of watch collecting is learning the correct terminology, the special words used in watchmaking that you have to know. This is complicated because different countries use different terms.

For example, hairspring, balance spring, and spiral all mean the same thing and are used in America, England, and French-speaking countries, respectively. Likewise pillar plate, bottom plate, and dial plate are terms used to describe the same part.

More confusing are these terms: movement in the gray, *ébauche*, and *mouvement en blanc*. Movement in the gray is an English term; it refers specifically to a watch movement with brass plates, before it has been gilded. And mouvement en blanc is the equivalent French term. The French word ébauche is confusing! Strictly speaking it refers to an unfinished movement, but just how much is unfinished is not clear. In the past an ébauche was a rough movement needing a lot of work before it would run. More recently the word has been used to describe completely finished movements that only require some engraving and beautifying. I will use the word ébauche to mean a supplied movement to be finished.

And some terms are simply strange! In the future, if you come across a part that is *à fleur,* it does not mean there is a flower in your watch; it means flush with. And in some watches you will find a "surprise piece," although I have never discovered why it is surprising. Most words are sensible, but a number, like surprise piece, are descriptive. So a snap is a groove and a click is a pawl, both named after the sounds they make.

Sensible or not, it is necessary that you learn these terms so that you can talk and write about watches correctly and be understood, so that you are not embarrassed, as you would be if you said your watch had a lid when you should say your watchcase has a cover.

Money

Despite what dealers tell us, collecting has very little to do with money. Sure, how much free cash you have will influence what you can buy and so what you collect. And most people, including me, occasionally have pangs of envy when we see others able to afford what we cannot. But this is generally a minor problem because we quickly adapt our interests and our desires to what is within our reach. This is why so many people collect Rolex watches: They are not too dear, there are plenty of them, and they are very good. We are happy to leave Patek Philippe watches to our more wealthy brethren.

Anyway, what watches cost depends on why we are collecting them. The famous Paul Chamberlain, who wrote *It's About Time*, collected watch movements without cases; he was interested in how they worked rather than how beautiful they were. And many of my watches have cost $100 or less. But if your heart's desire is a grand complication, then a spare million dollars may be necessary.

Not only do we buy to suit our pockets but we buy, as I have said, with little intention of selling. Dealers will talk about investments and profit, but it really doesn't matter to the dedicated collector. Oh, we will pay lip service to it, if only to tell our spouses that the money is as safe as in a bank, but it doesn't affect our decisions much. Indeed, if you buy as an investment, you are nothing more than a closet dealer, in which case this book is not for you, and I hope you have better luck than most people have with shares, horse races, and collectibles.

Over the years I have become increasingly annoyed by price guides and authors who write about investments. Such books are not for collectors; they are for people who think they can make money from watches, for amateur investors. Sometimes I think these books are produced by professional dealers as advertising aids. They always begin by showing the amateur how he or she can sometimes make amazing profits by buying and selling watches, carefully ignoring the fundamental reality that it is virtually impossible to buy from a dealer and make a profit. Indeed, it is extremely hard to profit no matter where you buy unless you are clever and knowledgeable enough to use the same techniques that dealers use. And that means becoming a dealer yourself, which is not possible for most of us because it is a full-time occupation.

Chapter 2

The Looking Game

Where to Start

As soon as possible after you have started collecting watches—hopefully before you pay a large amount for one—you should begin to study them. There are many things you need to know: history, different types of mechanisms, dating, and assessment of quality and condition, to name a few. But probably the best place to begin is by learning about styles.

The aim of learning about styles is to reach the point where, after a quick glance at a watch, you can estimate how old it is, where it was made, and how good it is. To be able to do this, you have to look at the cases and movements of lots and lots of watches—far more than you own or ever will own.

Learning about pocket watch styles is both easier and harder than learning about wristwatch styles.

First, it is easier because most pocket watchcases are simple to open so that you can see the works, the *movement*. Because they are much bigger than wristwatches, it is easier to examine them with a small magnifying glass and a few simple tools. In contrast, wristwatch cases often cannot be opened without special tools, and examining a tiny movement can be much harder.

Second, it is harder because pocket watches have been made for over 500 years in many different countries. As a result, there is much more variation in their design and more to learn. But you will discover that all these variations fit fairly neatly into a few styles adopted in different places and at different times.

Even if you want to collect wristwatches, I strongly recommend you start by examining pocket watches. It is quite likely that you will damage a watch when you first start to learn about it; the owner is not going to be pleased and you are going to be embarrassed or, worse, out of pocket. So a fundamental rule is: when in doubt, don't! The best way to avoid such unpleasantness is to start with easy tasks and never attempt anything unless you are certain you know what you are doing. And starting with large, more robust pocket watches will be a great help toward achieving that goal.

Another reason for starting with pocket watches is that nearly everything you find in a wristwatch was previously used in a pocket watch. So everything you learn about pocket watches will be useful for your real passion.

Look, Listen, Feel, and Smell

This is a self-administered quiz. The basic idea is to see how long it takes you to estimate how old a watch is, where it was made, and how good it is, and then to see if you are right. When you begin, you will have no idea how to do this, and your success rate will be very poor. But after a while you will be able to look along a row of watches in a display case and mentally categorize them into "not interesting," "interesting," and "don't know," and have a good idea of age and where they were made before you even pick them up.

Dealers and shop owners are going to hate me for what I am now going to tell you to do, and they also will soon dislike you if you are not careful. So there is one fundamental rule: Always be polite and considerate. I have been playing this game for about 20 years and I am pretty knowledgeable, maybe an expert. But I still stick to this rule, even if I know the dealer is ignorant. Only once did I forget and now that dealer spurns me. Being impolite, arrogant, or superior simply isn't worth it.

The second rule is: Only examine watches that indicate the time and nothing else. Watches with complications (repeaters, chronographs, even calendar and automatic watches) are simply too complex for a beginner to assess. By all means look at such watches, but just out of curiosity and not as part of the looking game.

The third rule is to say to yourself repeatedly and often, "Be polite; don't take risks; if in doubt, don't." You simply cannot afford to be overconfident.

To play the game you will need fingernails that are not too short and a tool kit that will fit in a small container that can be carried in your pocket (Figure 2-1).

Now that we are ready, go to a local secondhand dealer, antique shop, watch dealer, or auction house where there may be some watches to look at. Or go to a National Association of Watch and Clock Collectors (NAWCC) regional meeting or an antique fair. If it is an auction house, then you must be there at a viewing time well before the auction begins. Do not forget the object of the exercise is to learn, not to buy. If you think you might weaken, leave your money and credit cards at home!

Once in front of some watches, point at one and ask politely if you can look at it. Then take out your tool kit. The person behind the counter may show signs of panic or become reluctant. This is not surprising; if you damage a watch while examining it, the seller is going to lose money or force you to buy it. The less the person behind the counter knows, the easier it will be. A specialist watch

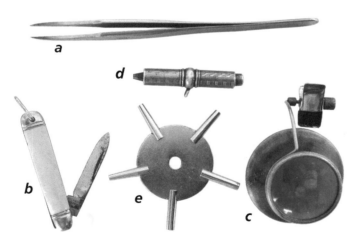

Figure 2-1.

1. A pair of fine tweezers **a**. Tweezers are very good for damaging watch movements! So don't use them unless you have to and know how.

2. A small, blunt pocket knife, **b**, or a proper case opener if you can get one.

3. A low-power loupe, **c**, a magnifying glass that fits in your eye socket or can be attached to your spectacles so that both your hands are free; holding a watch, a magnifying glass, and tweezers all at once is impossible.

4. If possible, an adjustable watch key, **d**, for key-wound watches. The one I have is spring loaded and opens up when you press the button on the end. Alternatively, get a bench key, **e**, with five keys radiating from the hub. It is limited to five different sizes and is not as easy to use, but it will do.

5. A pocket-sized English silver hallmarks book. I use Judith Banister's *English Silver Hall-Marks*, but there are plenty of others. Gold marks are basically the same and so another book is not needed for them.

dealer will be the hardest person to persuade. Remember the rules and when in doubt ask the owner to open the case for you so that you can look at the movement. Do not call it a "mechanism" or "the works"; use the correct terminology; at this point the only words you need to know are "watch," "case," "movement," and "please." If there are no objections, open the case yourself, if you can! Again, if in doubt, don't! If the owner refuses to allow you to open the case or open it himself, politely leave and go somewhere else.

During this process be very aware of the seller. If he seems to be knowledgeable, listen and try to remember what he says. There is nothing wrong with admitting you are ignorant if you see that he will be happy to help you and perhaps teach you a few things.

Having looked at the case and movement of one watch, close it up carefully, hand it back, and look at another. If the owner is happy and willing, go through as many watches as you can. Quite often the owner will eventually get a bit grumpy because you are not buying anything, in which case call it quits and leave.

If you can, give the shopkeeper some indication of your interest in particular watches. You might buy one (try not to) or you might come back later and buy one. After all, you are a collector and this is a local shop. However, buying the first watch you see is almost always a bad idea.

There is one other essential rule for this game and all the other exercises in this book: look at, listen to, feel, and smell watches! Right from the beginning you should start developing all your senses. Looking is the obvious and primary focus, but the other three senses are also vital.

Listen to watches. Each watch will make its own noises, mainly the ticking, as it marks the passing of seconds, but also other noises, like the sound it makes when you wind it, the sound when you set the hands, and so on. As you listen to more and more watches, you will come to distinguish the sounds of a watch in good condition from one that is sick.

Touch watches. Let your fingers explore the case and bracelet and feel how well covers close and open, and how the winding crown moves, and so on.

And smell watches. Clean watches smell clean, and dirty watches have a very distinctive smell. I am not sure if it is stale oil, slight corrosion, or bits of skin and dirt from the previous owner—probably a combination of all three—but after you have sniffed some dirty watches, you will remember it.

Cultivate these senses and you will be surprised at how much you can learn about a watch with your eyes shut.

Note that I don't mention museums. Although museums are very good places to learn about clocks, especially tower clocks, they are nearly useless when it comes to watches. This is because watches are locked away in display cabinets, and you cannot handle them, let alone examine them. So all you can do is admire the dial and case, and the few times when a movement is shown it is too far away from your eyes to be useful. Museums can be useful if you know a curator and can get behind closed doors. I have been to the famous Musée International d'Horlogerie in Chaux-de-Fonds, Switzerland, and it took me less than an hour to wander around looking at the watches on display, learning only a little. (Admittedly, it would take a beginner much longer.) But later I was fortunate enough to spend several hours in the restoration workshop examining just two fascinating watches.

A Few Examples, Mainly Cases

Pocket Watches

Now that you understand the game, it is time to learn how to play it. We are going to look at a few watches to show you how to look at them and what can be learned.

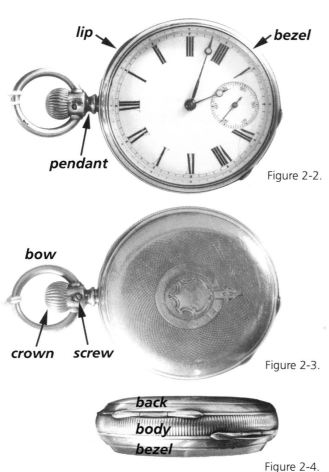

Figure 2-2.

Figure 2-3.

Figure 2-4.

Looking at the case of the first *open-face* watch (Figure 2-2) tells us very little. The case is silver. Silver has a distinctive feel and appearance and, because it is relatively cheap, such cases are almost always solid silver. There is a screwhead on the *pendant*, between the *crown* and the body of the case. The small seconds hand and its subdial are sunk below the main part of the dial—a *single-sunk dial*. In this watch the small seconds dial is a separate piece that is glued or soldered to the main dial. Although it is difficult to see in the photograph, the back is *engine turned*; this fine pattern of interlocking lines is a very common decoration (Figure 2-3). I suspect some decorations may be used to hide blemishes. It is extremely difficult to produce a perfectly flat, perfectly polished surface, and decoration tends to hide small faults. This watch has a thick *crystal* made of glass with a beveled edge.

Most important, at this stage, are the hinges or *joints*, shown in Figure 2-4. These joints are like any ordinary hinge, consisting of tubes with a pin running through them. (The correct term for a case hinge is joint, but for convenience I use the word hinge.) Both the *bezel* holding the *crystal* and the back are hinged to the case body. Crystal is an odd word to use, but the better word glass has to be avoided because some crystals are made of plastic. Old watches should have glass crystals.

Where there is a hinge there is something that opens! In the dial view you can see a lip on the bezel opposite the hinge. This lip is there so that you can put a fingernail behind it and open the bezel; there is also a lip on the back. The bezel has a small ridge around the edge that fits into a small hollow on the case body; this is called a *snap*. These snaps hold the bezel and back shut. Snaps are light when something needs to be opened regularly. Only use a case opener or a pocket knife if you cannot open a case with your fingernails. Put the blade under the lip and carefully twist it. Note that horology uses some common English words, but they have precise, special meanings that need to be understood. Snap is one and crystal is another.

The first thing to do is to open the back, where we find an inner cover, the *dome*. In this case the dome is hinged and has a small gap near the pendant. The snap is much firmer because normally the dome is only opened to repair the watch. Often the area around the gap will be scratched by previous attempts to open it with a knife blade (Figure 2-5).

On the inside of the back there are *hallmarks* (Figure 2-6). These hallmarks tell us that the case was made in England, it is sterling silver (the English name for 0.925 or 92.5 percent silver), the case was assayed (tested to make sure it is sterling silver) in Chester, and it was made between 1879 and early 1880 (the date letters cover a period from about March one year to March the next year). The initials "AB" tell us that the casemaker was Alfred Bedford. (How I know this is explained in chapter 3.)

Now open the dome. To do this you may need a case opener, which is usually decided after you have broken your fingernail. At last, inside the dome we find the movement (Figure 2-7).

Ah! it is an American watch in an English case! It was made by the American Watch Co. Waltham, MA (everyone calls the company Waltham), the movement design (which Americans call a *model* and others a *caliber*) was given the name Hillside, and the serial number (S/N) of the movement is 1,386,863. By checking a table of Waltham serial numbers, we find that the movement was made between 1877 and 1883 (I will tell you about dating in chapter 3). If (and only if) the case is the original one, the hallmarks enable us to refine the date to between 1877 and 1880. American movements were made in a few standard sizes, and it is easy to swap cases. This case is probably original and the number 18072 on it is the case number. Some cases have the serial number of the movement stamped on them, but many do not.

Figure 2-5.

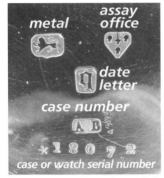

Figure 2-6.

What we are looking at is the *top plate* (Figure 2-7). This movement is a *three-quarter plate* movement where most of the wheels and other bits are under a disk of metal (often gilt brass, as in this case). At the top we can see the *balance cock* **a** and the *balance* **b** underneath it (the balance has small screws around its edge). Nearer the center we can see part of the *center wheel* **d**.

Underneath the dial is another plate, called the *pillar* or *bottom* plate. The top plate is screwed to three pillars on the pillar plate by the three blue-black screws around the edge of the top plate, and the balance cock is also screwed onto the pillar plate. All the wheels and other parts are inside the *frame* created by these two plates. The wheels **e** (Figure 2-8) are mounted on axles, which watchmakers call *arbors*, with a pivot at each end. The pivots sit in holes in the plates or cocks. You can see five circles **c** on the top plate with smaller circles inside them. The inner circles are the ends of the pivots. Around each hole there is a small depression, the *oil sink*, which holds oil to lubricate the bearing. The large pivot toward the bottom right of Figure 2-7 is on the end of the *barrel arbor*. The barrel contains the mainspring, which provides the motive power.

We have looked at the movement first because its quality and condition are the main factors in deciding if the watch is interesting and if we may want to buy it. If we are still interested, we had better see if it works.

Whenever you see a watch with a crown, the watch is pendant wound; turn the crown and you will wind

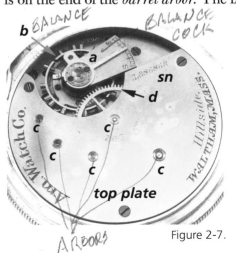

Figure 2-7.

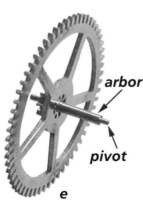

Figure 2-8.

the watch and hopefully it will run. (Very occasionally you will find that someone has put a key-wound movement into a case with a crown. But such marriages are quite obvious.)

As well as winding the watch, you must also be able to set the hands to the right time, and this is a bit tricky. I suppose you wear a wristwatch and know that to set the hands you pull the crown out a little. Many pocket watches are the same, and trying this is a good starting point. But no, you cannot pull out the crown; or you pulled it so hard it came completely out of the watch, and the person behind the counter has started crying with rage. Remember, be gentle and if in doubt, don't! Some watches set hands by pushing the crown in, so try that. Nope.

At this point it is tempting to think the watch is broken, but it isn't. Remember that the bezel is hinged. OK, open it and, lo and behold!, you will see the end of a little lever against the edge of the dial between the numerals XII and I. Pull the lever out, as shown in Figure 2-9, turn the crown and the hands rotate. There is nothing wrong; it is a *lever-set* movement. Note also there is a hairline crack running from the subseconds dial to the edge near VIII, which confirms that this is an *enamel dial* (the dial is made by melting opaque white glass onto a copper or gold disk).

There are a few more things we can learn by reading reference books. Waltham and other American companies made watch movements, but they did not make watchcases; they sold movements that were later put into cases and sold as complete watches. Waltham had a branch office in London, and movements were shipped to England, put into cases made in England, and sold or shipped out to the British colonies, including Australia. English casemakers had to be registered with the assay offices, so Waltham used the services of Alfred Bedford to get cases made and hallmarked in England. As a result of this arrangement, nearly all the American watches found in Australia, which is where I live, are made by Waltham and in English cases.

Before I forget, the screw in the pendant below the crown (see Figure 2-3, page 12) holds the crown on. Loosen it and the crown will come out with the *stem* attached to it (Figure 2-10). The screw runs in the groove in the stem so that the crown and stem can be rotated but not pulled out or pushed in. It is the stem that connects the crown to the movement. The stem end has a square hollow that fits over a square peg attached to the winding and setting mechanism. It is essential that the crown and stem can be moved out of the movement. If this were not possible, the movement could not be removed from the case to clean and repair it. Indeed, it would have been impossible to get the movement in the case in the first place! Nearly all crowns and stems have to be completely removed, but later we will see a watch where this is not the case.

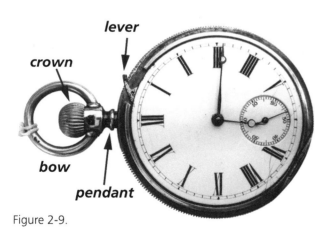

Figure 2-9.

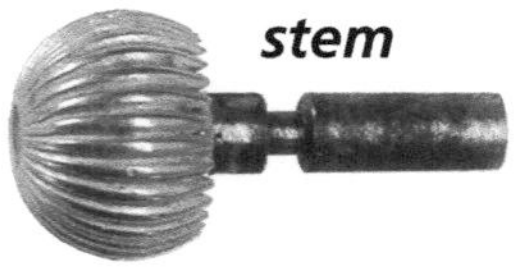

Figure 2-10.

Crowns and stems are attached in all sorts of ways, and we are not concerned with the details at the moment. Just remember, pull too hard and the crown and stem might come off completely, which is very embarrassing.

The following (Figures 2-11 to 2-17) is a second watch for our examination:

We know who sold this watch because the dial reads "G. Spiegelhalter & C°, Whitechapel Road London," but he may not have made it. This dial is also single sunk, but this time the seconds dial is an integral part of the main dial and not a separate piece. Again there are two hinges, so both the bezel and back open, but there is no lip or gap on the edge of the back to get a purchase with a fingernail or knife blade. There also is no crown to turn, so it is probably a key-wound watch.

How do we get in the back?

Figure 2-11.

Figure 2-12.

You are going to learn something now that sets you apart from other novice collectors. The back is spring loaded and held shut by a spring-loaded catch that fits under the snap on the back (Figure 2-12). To open the back, you press down on the *push-piece* in the pendant. Sometimes the cover will not open easily; it can be difficult to press down the push-piece far enough, or the spring that opens the cover may have broken. Some people also just pry the back open, which must not be done.

So far you are indistinguishable from other novices. The difference appears when you close the back. If you just push it closed until you hear a click, you are making a grave mistake. Every time this is done, the steel catch wears away the snap, as shown in Figure 2-13, cutting a big groove in it, until the case will not stay closed. The only solution is to get a silver-smith to repair the snap. To close any cover held by a catch, you absolutely must depress the push-piece, close the cover, and then release the push-piece.

Before looking inside this watch, I need to mention a problem with *hunter* watches; one is illustrated later in this book. A hunter watch has a hinged cover over the crystal, and this cover is opened by pressing on the push-piece or the crown; as a rule the push-piece or crown opens the cover that is opened most often, in this case the cover over the dial, so that you can read the time.

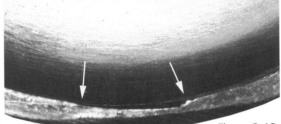

Figure 2-13.

There is very little space between the hunter cover and the bezel that holds the crystal, so the crystal is always very thin, and if it is made of glass, it can be cracked or broken very easily. So if you are examining a hunter watch, it is essential that you handle it very carefully, especially if you have to open or remove the bezel for some reason, as we had to in the first watch we looked at.

OK, now open the back, and we see a dome with two holes for keys: one to wind and one to set the hands; see Figure 2-12. Usually, but (to make your life difficult!) not always, the hole in the center sets the hands. Next, if you look at the inside of the back (Figure 2-14), there are no hallmarks. There is an "S" in an oval, which presumably means silver (from the feel and look of the case, it definitely is), the initials C. G, and the number 9608. So,

Figure 2-14.

despite the name on the dial, it looks like the case was not made in England.

Now to open the dome and look at the movement. And here we find a problem: no hinge! no nothing! In fact, the dome is a fixed part of the case and cannot be opened without a hacksaw. Some people get stuck in their thinking and, having learned to open the dome on one watch, try to open every dome. The worst case I have seen was after someone had used a large screwdriver and a hammer to punch a hole through it!

Figure 2-15.

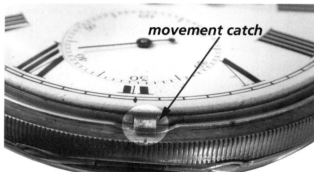

Figure 2-16.

Figure 2-17.

All right, we know the bezel opens. And under the bezel we see two things (Figure 2-15):

Near the pendant there is a hinge and under the dial on the opposite side there is a catch (Figure 2-16). Press in the catch and the whole movement and dial can be turned up out of the dome (Figure 2-17). This is not quite as easy as it sounds. The catch is a tiny block of steel held in place by a spring. It has a narrow slot in it for the end of your fingernail (which hopefully you haven't broken), and you have to simultaneously push in and lift. It is very easy for your nail to slip. Often nothing happens, but sometimes it does. You can bend one or both watch hands, or the minute hand can slip under your nail like a splinter of wood; I am sure you know just how painful that is! The worst thing that can happen is that you can chip the dial, which is a permanent and devaluing damage to be avoided at all cost, but is seen all too often.

Before looking at the movement, note that the bow is held onto the pendant by a screw that runs through the pendant and the push-piece (Figure 2-15).

This movement (Figure 2-17) is a bit like the Waltham movement, but it is best described as a *half-plate*. The terms three-quarter-plate and half-plate are used rather loosely. They don't refer to the physical size of the plates, but to the number of wheels held under them. You can see the top plate covers fewer wheels, and there are three cocks, a balance cock **a**, an escape-wheel cock **b** (under the balance), and a fourth-wheel cock **c**. At this point I should explain that generally a cock is a metal bar that supports one top pivot and is held in place by one screw. Again, the plate is screwed onto three pillars on the pillar plate. The balance is a plain ring of what looks like a low-grade gold, but it may be gilt brass.

This movement is inscribed "G. Spiegelhalter & Cº. Whitechapel Road London 9608." The number 9608 is the serial number and because this number is engraved on the case, we can be pretty sure the case is the original one and has not been replaced.

By the way, if you see a watch movement with a number like 1932 engraved on it, do not think this is a date. The number is invariably a serial number.

The absence of hallmarks makes it difficult to date this watch and suggests the case and movement were not made in England. The small steel cap (highlighted) that supports the winding square is typically Swiss. One reference book notes that Spiegelhalter worked between 1851 and 1881, so we can have a wild guess and say it was made circa 1865.

Having looked at these two watches we can pass over the next two examples quickly and just look at the main points.

The case of the next watch (Figures 2-18 to 2-21) is basically the same as the first watch we examined, except that only the back has a hinge and the bezel is not meant to be opened, which is OK, because the hands are set by pulling out the crown. The case is hallmarked, and the marks are for sterling silver, the Birmingham assay office, and the date letter for 1906-07. The casemaker is "ALD." Around the casemaker's mark there are a number of small, scratched letters and numbers. These are repair marks. Sometimes when a watch is repaired, the repairer will scratch information on the case so that if the watch is returned to him, he will know what he had done previously.

Although the movement is only engraved with "English Made" (Figure 2-20), the dial tells us the maker (Figure 2-21). The serial number is 819,275. Again we have a three-quarter plate movement with separate balance cock. The obvious differences from the Waltham movement are the two large wheels with screws in their centers. These wheels are the winding mechanism and are connected to the crown.

The Lancashire Watch Company attempted to make watches by machinery, but it failed about four years after this watch was made. The casemaker is A. L. Dennison, the famous American who helped found the Waltham Watch Company and who later emigrated to England and established a large watch-case factory.

Figure 2-18.

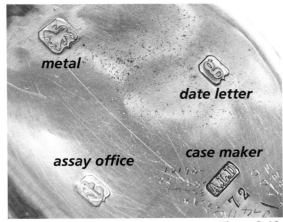

metal

date letter

assay office

case maker

Figure 2-19.

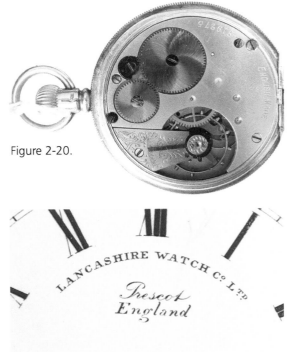

Figure 2-20.

Figure 2-21.

push-piece

Figure 2-22.

Figure 2-23.

Figure 2-24.

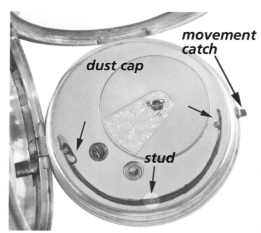

movement catch

dust cap

stud

Figure 2-25.

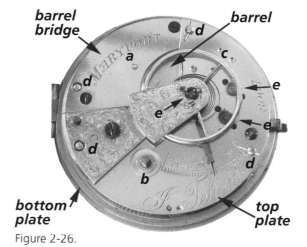

barrel bridge

barrel

bottom plate

top plate

Figure 2-26.

The fourth watch (Figure 2-22 to 2-27) is an English watch made by J. Telford, Maryport, and the case is very similar to the one for the Spiegelhalter watch. Both bezel and back are hinged and there is a push-piece to open the back. If you are not sure which cover a push-piece opens, it is usually the cover that is most often opened, in this case the back to wind the watch. Once again the inner dome is fixed, but we immediately see one difference from the Spiegelhalter watch; there is only one keyhole, which is for winding. To set the hands, the bezel must be opened and a key put on the square in the center above the hour and minute hands (Figure 2-24).

Like the Spiegelhalter watch, the movement is hinged to the case and we access it by opening the bezel and pushing the movement catch, which you can see in Figure 2-25 at the right-hand edge of the movement. When we lift up the movement, we are confronted with a brass *dust cap* covering all the movement except part of the balance cock. Removing this dust cap is easy, but I have seen many that have been broken through ignorance and carelessness. The dust cap is held in place by a semicircular steel catch, marked here by the arrows. Screwed to the movement are two posts with notches that protrude through the dust cap. As shown in the photograph, the dust cap is locked in place by the ends of the catch that fit into these notches. To unlock it and remove it, the catch has to be rotated clockwise; for this there is a small stud in the middle of the catch for a fingernail. Some catches have to be turned counterclockwise.

Often you will find watches without dust caps. But if there are two steel posts with notches, then you know that there should be one. There are four points to note about this movement (Figure 2-26).

First, the top plate of this *full-plate* movement is pinned to the pillars on the bottom plate. At the two points **d,** on the right and at the top, you can see the ends of the pillars and the pins going through them. At the two points **d** on the left, one pin is covered by the *barrel* *bridge* and one by the balance cock. Note that a bridge supports one or more pivots and has a screw at either end, whereas a cock has only one screw. Note also that the pillars are not uniformly spaced around the edge of the plate.

Second, this is a jeweled movement. The three arrows **e** point to pivot holes that have jewels set into the top plate and the balance cock. Jewels with holes drilled through them are used because they are harder than brass; they reduce friction and wear, and so increase the useful life of the watch. All the watches we have looked at have had a jewel in the balance cock, but this is the first watch with plate jewels. At the moment I am discussing what we can see. In fact, this watch has 11 jewels, but only three are obvious.

Third, in Figure 2-26, **a** is the barrel arbor; you can see part of the barrel under the plain gold balance, and **b** is the winding square. A movement with a separate barrel arbor and winding square is almost always a *fusee movement*, which this side view below shows (Figure 2-27). Except for some very rare watches, fusee watches are always key wound and set. Movements without fusees are called *going barrel* movements.

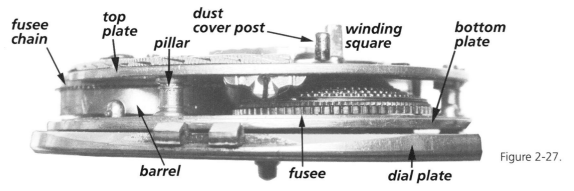

Figure 2-27.

Also note that this watch has three plates: a top plate, a bottom plate, and a separate *dial plate.* Many watches attach the dial directly to the bottom plate, but some use a dial plate to create space between the dial and the bottom plate for the motion work, the wheels that drive the hands.

Fourth, at **c** in Figure 2-26 you can see the ends of two small pins. These two *banking pins* tell us that the movement is almost certainly a lever movement using a lever escapement. What a *lever escapement* is doesn't matter now, but the telltale banking pins are a quick way of identifying such a movement.

There are some watches that seem to be fusee movements when you look at the top plate, but they are, in fact, *dummy fusee* movements; one is shown in Figures 2-28 and 2-29. If you compare it to the movement in Figure 2-26, there is an obvious difference. The winding square of the fusee watch is on the fusee arbor, but the winding square of the dummy fusee watch is on the barrel arbor. Sometimes a dummy fusee movement

Figure 2-28.

Figure 2-29.

looks as though it has two barrels, as in this case, and they are often incorrectly described as watches with two mainspring barrels. The barrel on the left in Figure 2-29 is the mainspring barrel, and the barrel on the right is a dummy barrel. The reason for these watches gives an insight into human nature. If you wind a going barrel watch, you turn the crown or the key clockwise, but to wind a fusee watch, you must turn the key counterclockwise. People living in England were so used to winding their watches counterclockwise that when going barrel watches were introduced, they had trouble winding them! So some going barrel watches were made with a dummy fusee (acting as an extra wheel) so that they could continue winding the way they were used to!

The next watch (Figures 2-30 to 2-36) has no hinges and no lips, although there is a raised rim all around the back and the bezel. Such watches need to be treated with care because the cases can open in two quite different ways.

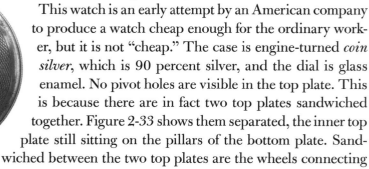

Figure 2-30.

Figure 2-31.

This particular watch is pendant wound and set (by pushing the crown in), and there is no need to open or remove anything except for repair. In this case the bezel and back snap on. Sometimes fingernail pressure is enough, but usually a case opener is needed, and only the back needs to be removed to inspect the movement. This movement and its case are made by the Waterbury Watch Co., and it is a Series K "Charles Benedict" model; again it is a three-quarter plate movement where everything other than the balance and one wheel is under the top plate. The hands are, in fact, luminous hands, but all the luminous material has been removed, leaving the fine, steel framework.

This watch is an early attempt by an American company to produce a watch cheap enough for the ordinary worker, but it is not "cheap." The case is engine-turned *coin silver*, which is 90 percent silver, and the dial is glass enamel. No pivot holes are visible in the top plate. This is because there are in fact two top plates sandwiched together. Figure 2-33 shows them separated, the inner top plate still sitting on the pillars of the bottom plate. Sandwiched between the two top plates are the wheels connecting to the crown for winding, which have been removed, and the holes for the pivots of the other wheels. The pillar plate is also a double plate, and this watch, although "cheap," is well constructed with a very interesting design.

Figure 2-32.

Figure 2-33.

Figure 2-34.

Figure 2-35.

Figure 2-36.

Although the purpose of this book is to teach you about watches, you may also become interested in associated things. This watch was acquired in its original cardboard box (Figures 2-34 to 2-36), which is in surprisingly good condition considering it is about 125 years old. Original boxes, keys, and other ephemera can add significantly to the collectibility of an item.

As I noted above, watchcases without hinges can open several ways. The Lemania Chronometre in Figures 2-37 to 2-40 also has a rim around the edge of the bezel and back, but this time both unscrew. It is vital that you check a watch very carefully before you try to pry off a cover; many otherwise nice watchcases have been ruined this way. The Lemania gives you a hint because the rims are milled to make it easier to grip them; in contrast, the rims on the Waterbury watch are smooth.

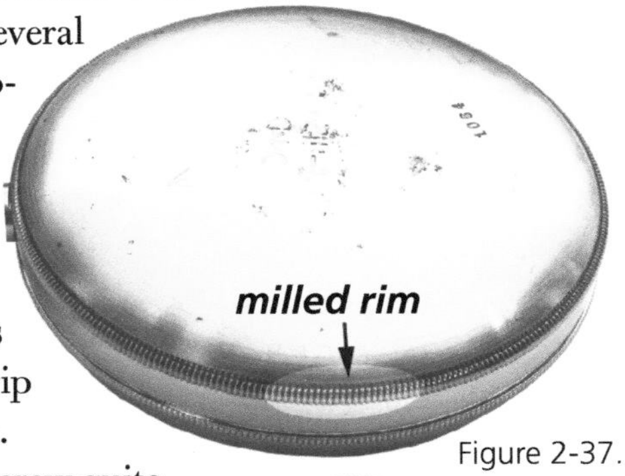

milled rim

Figure 2-37.

Pocket and wristwatch cases that screw together usually unscrew quite easily. The main problem is to screw them back together correctly without damaging the threads.

This Swiss-made watch is the best we have seen so far. It was used for navigation by the British defense forces and has the "broad arrow" military markings on the back; it comes with a rating certificate and its original box. The case is completely smooth, practical, and made of a base metal, probably nickel steel. We also can see five jewels: the three marked by arrows in Figure 2-40, the balance cock, and the other cock. (What looks like a black blob on the other cock is actually a piece of highly polished steel with a mirror finish. This is called a black polish because at some angles it appears black.)

Figure 2-38.

Figure 2-39.

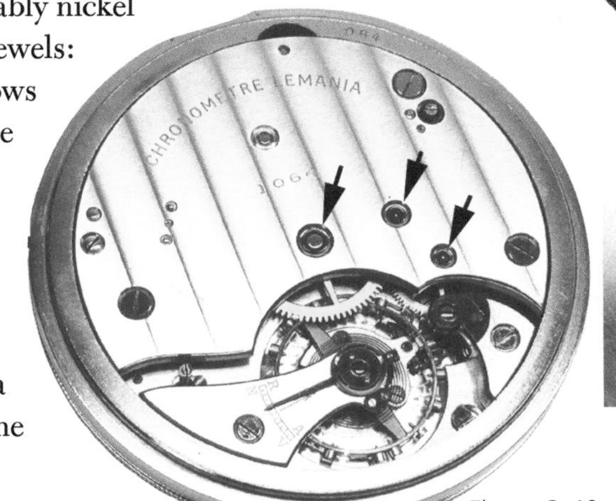

Figure 2-40.

chapter ring

Figure 2-41.

setting lever

hinge

Figure 2-42.

Figure 2-43.

Figure 2-44.

The last pocket watch shown here is a Hamilton watch made in America, which is also in a nickel case (Figures 2-41 to 2-44). The case is completely smooth, and the only indication of how to get into it is a milled band around the bezel and a faint line between the bezel and the rest of the case. There is no indication of how to get the back off—because it doesn't come off; the back and the body are a single piece. And because it is quite clear that there is no way to pry the bezel off, it must unscrew. But before doing so, note that the dial is *double sunk*. The center of the dial is a separate piece set below the outer *chapter ring* with the numerals, and the subseconds dial is sunk below both.

Once we have the bezel off, we can see two things. First, there is a hand-setting lever at the numeral 2 that pulls out; so the crown winds and sets the hands in the one position, like the first watch we examined. Second, there is a hinge and a nice, fingernail-shaped gap opposite it; this is beginning to look as though it is identical, in principle, to the Spiegelhalter and Telford cases we have already opened. OK, rest the watch in your palm with the pendant facing you, put your fingernail in the slot, and lift. The movement rises up about 5 mm and jams. If you are foolish, you will forget my rules and tug, which could damage the watch. What you should do is ask yourself (or the owner) "what is the problem?" The problem is that the stem attached to the crown fits into the movement and stops it swinging up. This is why on this watch the crown can be pulled out into what would be, on other watches, the hand-setting position, but isn't; it is pulled out into this second position to free the movement so that it can be lifted out of the case back. This stem is like that illustrated in Figure 2-3 (see page 12), but it is attached to the pendant in a different way.

The movement of this watch (Figures 2-43 and 2-44) is quite different from the others because it is simply beautiful! The warm glow of gold contrasts with the polished nickel of the full-plate movement and the artistic engraving overlaying the very fine, wavy lines. These wavy lines are called *damaskeening*. This is not the same as *damascening*, which you will find in dictionaries, but a term that is particular to American watches with this type of decoration. Two photographs are needed: Figure 2-43 shows the engraving and Figure 2-44 highlights the damaskeening. It is certainly well worth the effort to open the plain, rather uninteresting case.

However, beauty lies in the eyes of the beholder! American watch factories used to manufacture and sell movements, and entirely separate companies made cases. When you went into a shop to buy a watch, you would see a display case of movements and another display case of cases. You would pick out a movement and a case, and the watchmaker would put them together for you. So the beauty of this Hamilton watch is simply a sales aid. All American companies "dolled up" their movements in the hope you would find them more attractive than the "dolled up" movements of another maker.

This explains two things about American watches. First, in general, movements and cases can be swapped easily, because nearly all are made to standard sizes. And second, very high-quality movements, like the Hamilton model 940 illustrated, are sometimes found in cheap, plain cases. Indeed, "never judge a watch by its cover" is a basic rule in watch collecting.

One other thing to note is that we can see six jewels mounted in small disks of metal, called *chatons*, which are held in place by small screws; there are five in gold chatons on the plate and one on the balance cock. But the movement is engraved "21 jewels," so there are another 15 jewels somewhere.

In contrast, the Lemania in Figure 2-40 (page 21) was manufactured for military use, and the only thing that mattered was how well it worked. In terms of timekeeping, the Lemania and the Hamilton are about the same, and most of the appearance of the Hamilton watch is like makeup, which beautifies but doesn't alter the underlying watch or person.

This Hamilton watch is an *American railroad watch*. After a number of disastrous train crashes, the Americans set up standards for the design, quality, and accuracy of watches used by railroad companies. These watches are among the best for accuracy.

Wristwatches

In this section, to complete our exercise in examining watches we will look at wristwatches. All these examples are "junk" watches that I have accidentally collected or borrowed. Although they all work, they have faults that mean they are in rather poor condition and not really collectible. But they are very good for learning.

Early wristwatches usually had cases that snapped or pressed together. Opening them is the same as opening a pocket watch. Figures 2-45 to 2-47 show such a watch; both the bezel and the back snap onto the case body. Somewhere on the back and bezel there will be a small gap for an opening tool; often this is under the lugs for the bracelet or strap. The dial is made of metal with applied gold numerals slightly higher than the surface of the dial. The case back is stamped "Handly, Stg Sil"; Handly was a Swiss casemaking company. The other marks on the back are repair marks made by watchmakers. The bracelet is attached to the case by the wire lugs soldered on both sides. There is no maker's name on the movement and all we know is that it is probably a Swiss watch.

There are also similar wristwatch cases with hinges (joints), which are normally under one of the bracelet fittings.

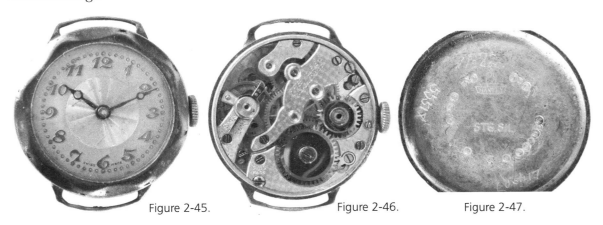

Figure 2-45.   Figure 2-46.   Figure 2-47.

23

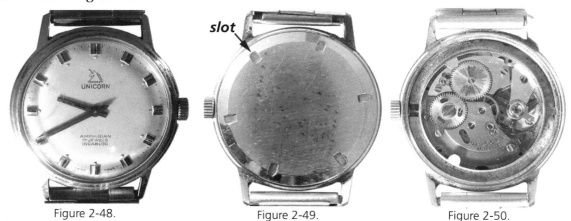

Figure 2-48.

Figure 2-49.

Figure 2-50.

slot

The second wristwatch is labeled "Unicorn" (Figures 2-48 to 2-50). The bezel and the case body are a single piece and the back screws off. To remove it a special tool is needed. One such tool is the "Jaxa" case opener shown in Figure 2-51. It is adjustable and has a number of sets of three interchangeable inserts to suit different cases. They fit into the slots in the case back; the Unicorn has six rectangular slots (Figure 2-49). When the back is removed, we see the movement is engraved "Unicorn 17 Jewels Swiss."

Figure 2-51.

The next watch (Figure 2-52) also has a screw back, but the Jaxa tool cannot be used to open it. This is a "Tudor" watch in a case stamped "Oyster Case by Rolex" and it is, in fact, a watch made by Rolex. The finely milled edge, which is inset in the case back (Figure 2-53), requires a special tool; there is no picture of the movement because I don't have the right tool to open the case. There are many other wristwatch cases that need different forms of spanner to open them and large numbers of pressed-metal spanners were made and distributed to watchmakers. However, nearly all can be opened with the Jaxa tool.

The marks on the back of the case shown in Figure 2-53 are caused by acids from the wrist seeping through the woven strap and corroding it.

Snap cases must be treated with care because not all can be closed easily, and it would be very embarrassing to have taken a back off only to be unable to put it back on! The next watch, signed Bucherer (Figures 2-54 to 2-56), is a good example. Instead of normal hands, the watch uses four concentric discs to display (from the inside) seconds, minutes, hours, and day of month (Figure 2-56). The move-

Figure 2-52.

milled edge

Figure 2-53.

ment looks good, but notice that it sits off-center in the case (Figure 2-55). This may indicate a problem or it may be intentional to position the display discs correctly (it is actually intentional). This automatic movement is marked "25 jewels, adjusted 3 positions" suggesting it is a high-quality movement. It has a typical semicircular rotor mounted in the center for automatic winding.

Figure 2-54.

Figure 2-55.

Figure 2-56.

Having examined the watch I tried to snap the back on and couldn't! The only way I could get it on was by using a small press made for doing this and for fitting crystals (Figure 2-57).

The last watch in this section (Figures 2-58 to 2-60) is a brand new Longines automatic; this is one of their "Le Grande Classique" range. It has a display back so that you can see the movement without opening the case and has probably been manufactured as a "collectible" watch.

A new watch can be collectible, but if that is your intent it should never be worn. Such a watch is generally only worth the current retail value, and as soon as you wear it, it becomes a secondhand watch and can lose much of its value. Worse, any scratches or other marks from opening the back to look at the movement will significantly degrade the watch in the eyes of the collector. My Longines watch is unused and kept with its original booklets and tags in its original box; it might appreciate in value in the future, but it might not! I don't collect wristwatches, so I could wear it. But I only wear cheap quartz watches, to avoid the risk of damaging a good watch, and so I expect it will spend its life in a drawer.

Figure 2-57.

The purpose of this section has been to show you how to open cases so that you can look at the movement. As you have seen, this is generally quite easy with pocket watches, but it can be very difficult with wristwatches. The wristwatch collector is often faced with the problem of having to decide whether or not to buy a particular watch without seeing inside. And wristwatch collectors are seriously hampered because I can open my pocket watches and examine their movements whenever I like, whereas getting into wristwatch cases is difficult and often cannot be done on the spur of the moment. So, to some extent wristwatch collectors are limited to judging by external appearances, and collecting must be based more on brands and information from books. But such a limitation does not mean the wristwatch collector can ignore studying the history and mechanics of timepieces.

Figure 2-58.

Figure 2-60.

Figure 2-59.

Signatures and Makers

We can now consider some other aspects of watches. Figures 2-61 to 2-64 show another Unicorn wristwatch. It is in a roughly rectangular case made by Handly, which is a two-part case with the front (the bezel and part of the case body) locking friction tight into the back. When it is prized apart, the front comes off, leaving the movement sitting friction tight in the back. It is usually quite easy to lift the movement from the back by the crown, but sometimes you will need to work it out using a knife blade between the dial and the back, taking great care not to bend or damage the dial. The movement itself is quite good, but unmarked except for the inscription "15 Jewels." The jewels appear to be in gold chatons pressed into the cocks and the bridge (Figure 2-64).

We have here a potentially nice watch, but who or what is Unicorn? If we have a knowledgeable friend, or we search through a few reference books (or the index in my book *Mechanical Watches*), we will learn that "Unicorn" is a trademark of the Rolex Watch Company; there never has been a Unicorn Watch Company. Well! A Rolex watch, that must be interesting!

Now normally the matter might rest there, but I happen to know that this is not the end of the story. To check my ideas I removed the dial to see what was underneath (Figure 2-65). As you can see in the enlargement of the outside of the bottom plate (the pillar plate), there are the letters AS in a squiggly shield and the number 984 (Figure 2-66). The letters AS are the identification mark of A. Schild S.A., and the number 948 is the *caliber number* that identifies this particular movement design. Schild was part of the Ebauches S.A. umbrella company. At various times the watch industry in Switzerland, with the help of the Swiss government, organized and reorganized watchmaking companies into cooperative groups; Ebauches S.A. was one such group, and S.A. stands for Société Anonyme, the Swiss designation for an incorporated company. These days, many Swiss watch manufacturers operate within privately owned groups, such as the Swatch Group. Schild and the other members of Ebauches S.A. made watch movements, which are called ébauches, and sold them to other watchmaking companies. So this Unicorn watch made by Rolex

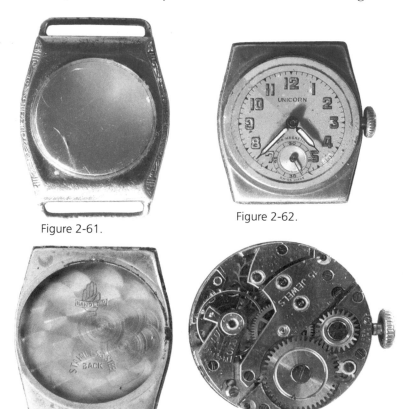

Figure 2-61.

Figure 2-62.

Figure 2-63.

Figure 2-64.

Figure 2-65.

rust

Figure 2-66.

uses an A. Schild ébauche put in a case made by another company, Handly! It seems Rolex had very little to do with it.

Four names, one watch, so who actually made it?

The problem of names and who actually made a watch is very important, and it is the reason why a study of the history of watch manufacturing is essential for collectors. In the process we will discover that Rolex produced watches with the signatures "Unicorn" and "Tudor." The company reserves the signature "Rolex" for the best watches it produces, but it also produces watches that use ébauches of a lesser, but still very good, standard. These watches carry the "Unicorn" and "Tudor" brand names. Actually, for a long time all Rolex watches were based on ébauches; the movements signed "Rolex" were sourced from a Swiss company called Aegler. See Dowling and Hess's *The Best of Times—Rolex Wristwatches: An Unauthorized History*. We will also discover that the Waltham Watch Company in America produced some cheap movements that it thought might tarnish the good name of the company; so these watches were just engraved "Home" and the company name was omitted.

Before leaving this watch, look carefully at Figures 2-61 to 2-66. Except for cleaning and a new crystal, this watch seems to be interesting and in very good condition. But if you were able to take off the dial, you would discover serious rusting near the crown (Figure 2-65). Often collectors face the problem of assessing watches without being able to check everything, and even the most immaculate looking watch may hide an unpleasant surprise.

Figure 2-67.

Another example is shown in Figures 2-67 to 2-69. The dial is signed "Jules Jurgensen since 1740" and the movement, which is very small, is stamped "J. Jurgensen" and "Seventeen jewels." Despite the elegant dial, the back of the case is worn, showing the "gold" to be a very thin coating over a steel case. In addition, the movement looks ordinary. Hiding under the balance is a very small stamp. The number 89-21 is a caliber number and the trademark is for the Ebauches S.A. company Fabrique d'Horlogerie de Fontainemelon (FHF) type standard (ST). This is confirmed by the official catalog of Swiss watch repair parts, which shows an almost identical caliber 69-21 made by that company. The date of that watch is 1973, and the Jurgensen watch illustrated above has an inscription that dates it to 1977.

Figure 2-68.

The Jurgensens were a family of Danish watchmakers who produced superb work that is very collectible. But it is clear that this is a very ordinary watch trying to gain status by using that name.

You can also see a metal clip over the balance jewel, marked by the arrow (Figure 2-69). This U-shaped spring is a shockproof system. The balance jewel is loose and held in place by this clip. If the watch is dropped, the jewel can move and so avoid breaking the very fragile balance arbor, the balance staff. I know this watch has been repaired by someone who was careless. One end of the clip is visible at the upper right, but it should be tucked under the metal ring just below it; the repairer has failed to twist the clip around far enough when he put it back on.

Similarly, my Longines wristwatch in Figure 2-57, page 25, uses an ébauche made by ETA, which supplies movements to several Swiss watch companies.

Figure 2-69.

Another example of this problem of determining true origins is the G. Spiegelhalter & Cº. watch I looked at in this chapter (Figures 2-11 to 2-17, pp. 15-16). Although I didn't comment at the time, the case is a typical nineteenth-century English case; hinged back and bezel, fixed inner dome, the movement hinged and lifting out to the front. Almost every nineteenth-century English watch is in an identical case. But it is not an English case because it does not have English hallmarks, and the movement has features that strongly suggest it was made in Switzerland! Although with this watch I cannot be certain, I am very strongly of the opinion that none of it was made in England. So was George Spiegelhalter a watchmaker or something else? And do the names we see on watches mean anything at all?

In contrast, the Telford watch (Figures 2-22 to 2-27, pp. 18-19) has a typical English movement and everything indicates that the whole watch was made in England. But, just as with the Unicorn watch, many English pocket watches have initials and numbers stamped on the pillar plate under the dial.

Figure 2-70 shows the outside of the pillar plate, normally hidden by the dial, of a movement made by Newsome for the watchmaker whose name appears on it. The small numbers marked by arrows are its size, the diameter, and pillar height. England also had some ébauche makers and watch factories (but in England such movements are called "movements in the gray") and many watches with names on them actually came from factories such as Newsome, John Wycherley, and Rotherhams.

Figure 2-70.

In order to make some sense of this, the obvious, visible name on a watch is called the *signature*. Often it is on the dial, but sometimes it is only on the movement. The person or company whose signature appears on a watch may or may not be the maker, the company, or individual who actually constructed and assembled the parts.

Some watches are easy. I have several pocket watches signed "Bushman made for Taylor & Sharp Hobart" (Hobart is where I live in Australia). This is clearly a signature, which the movement confirms because it is stamped "Tavannes," the name of a Swiss watchmaking company. Anyway, I know that there never has been a watchmaking company in Australia and all watches signed with Australian names are imported. Taylor and Sharp, a local jeweler, would have placed orders with the Tavannes company for watches with their signature on the dial, and all they did was sell them. It is highly likely that this is the case with Spiegelhalter as well.

In some situations it is hard to decide who should be called the maker. Some very famous brands, including Rolex and Patek Philippe, use movements made by Valjoux, a Swiss ébauche manufacturer. But often these movements undergo extensive adjusting and finishing to bring them to the high standard required by Rolex and Patek Philippe. Does this difficult handwork, assembling, and finishing components mean we can regard Rolex and Patek Philippe as the makers?

Over the years the habit has grown up of calling anyone whose name appears on a watch a *watchmaker* and anyone who sells or repairs watches also a *watchmaker*. Even today watches are taken to watchmakers for repair, but probably nobody behind the counter has ever made a watch and never will; quite a few can only change batteries and adjust bracelets. We are stuck with this carelessness with words and it is too late to change. So, as much as I don't like doing so, I will simply talk about the maker of a watch without distinguishing between maker and signatory except when we need to do so.

The collector of American watches has a much easier time. Except for private label watches, where the name on the watch is that of the retailer and not the maker, and fakes, which I will discuss later, the vast majority of American watches are signed by the company that actually made them.

But What about Styles?

If you remember the beginning of this chapter, you will recall that I talked about styles, and since then I have completely ignored the topic! Actually, nearly everything I have covered does relate to styles.

My main aim has been to teach you how to successfully get into (and out of!) a watchcase with a nervous dealer looking over your shoulder. And don't forget my examples are just that, examples. You have to go out and do it yourself. And keep doing it for as long as you collect watches.

When you do so, you will quickly find that the examples I have described are representative of different case styles. And you will also discover that those case styles are quite closely related to signatures, makers, and, consequently, countries. For example, I have mentioned that the Spiegelhalter and Telford cases are typically English. As you examine more watches, you will find that such cases almost always have English hallmarks and almost always have English, Irish, or Scottish signatures. Occasionally, you might find a watch in such a case with an American or Canadian signature; but I will bet that further investigation will teach you that behind nearly all these signatures lurks an English maker!

One case style I have not illustrated is the *pair case*. The name means two cases, so do not call it a *pear case* as some people do. As a rough rule, such cases were made before 1850, although there are some with later dates. The outer case is held shut by a spring catch operated by a push-piece and the inner case is like the Spiegelhalter or Telford case without the outer back cover, just the dome with the winding hole.

Likewise, if you examine wristwatch cases, you will see some definite trends in shape, construction, and materials, although these styles are less clear than those in pocket watches. They also are further complicated by watch companies producing retrospective series, new watches in old styles to appeal to collectors and the dictates of fashion.

While you are playing the looking game, you will become more and more aware of movement styles. One thing that will become clear is that the Telford watch has a typical nineteenth-century English movement. Recognizing this style can be very useful. There are many watches signed "Tobias, Liverpool" to be found in America (or, if you live elsewhere, on the eBay Internet auction site). You will find a few that are typically English, and you will find many more that are not. Even though the following watch (Figures 2-71 to 2-75) was not made by Tobias, it illustrates the point.

This watch has a *hunter case* with a cover over the dial. The hunter cover usually has a shield for

Figure 2-71.

Figure 2-72.

Figure 2-73.

the owner's monogram or some other decoration to distinguish it from the back cover. The push-piece in the pendant opens the hunter cover and must be depressed when closing it. Always look carefully and learn what you can at each step. Here we can see that the push-piece is unusual—it looks a bit like a hat—and when we press it, the hunter cover does not spring open; the catch works but the *secret spring* is broken. (Grandparents used to like astonishing their grandchildren by holding their watch in one hand, incanting magic words, and then the hunter cover would suddenly pop open! Hence the term secret spring.) Also note that the watch has been well used; the case is in good condition but very worn, and much of the shield has been rubbed off.

The inside of the hunter cover is stamped "Warranted Coin Silver 19715." The inside of the back cover has the same number and what appear to be hallmarks, shown in Figure 2-73. When you have looked at enough English hallmarks, you will have no trouble in realizing that the symbols on this case are not hallmarks. In addition, all English silver cases are sterling silver, never coin silver. But coin silver, which is 0.900 or 90 percent silver, is a common grade used for American watch cases and the marks are American.

Because there is no information about the maker, we need to attack the back. We have no idea how to get at the movement, but when in doubt open the back to see if the dome is fixed or hinged.

Ah! The dome is hinged and it is replete with information: "Straight line lever. Full jewelled. Hands. No. 19715. J.A. Hamann, New York." and the movement is also inscribed "J.A. Hamann, New York."

There are a couple of useful points to note. First, Swiss watches are often inscribed like this, but usually in French. Consequently, a lot of people believe such watches are made by Aiguilles, which, sad to say, is nonsense; *aiguilles* is the French word for hands. The Swiss are just being helpful by identifying each key square. Second, the movement and the case clearly belong to each other because both have the same serial number.

Most importantly, the movement is a typical Swiss *bar movement* of better than average quality. There is no top plate; instead, there are five cocks and two bridges. For convenience the term top plate is still used to distinguish this side of the movement from the dial or bottom side. The Swiss made millions of bar movements, and you will come across many of them.

Figure 2-74.

Figure 2-75.

Is it an American watch? Almost certainly not, because the movement and the inscription on the dome are typically Swiss. However, the term coin silver and the marks on the case suggest the case was made in America! So Hamann probably imported movements that were cased in America.

To go back to Tobias of Liverpool, you will find many Tobias watches are like the Hamann watch and typically Swiss. Indeed, even without reading much, you might deduce that a lot of the Tobias company business was with exports to the United States and, after exporting some English-made movements, the company switched to importing Swiss watches (complete in cases), signing them, and re-exporting them to America. But the story is more complicated than that, as I explain later.

What To Do When the Shops Are Shut

Because my main aim in this chapter is to get you to go out and look at watches, I have not yet asked you to read anything. One reason for this is that, as a beginner, you probably don't have much of an idea about what you want to collect, and there is not much point buying and reading books if, later, they simply end up gathering dust. And the second reason is that I want to direct you to good, meaty books. Coffee table books full of color photographs are fun, and some are useful, but most are of little real value to the serious collector. If you have the collecting bug, then you should be opening watches, looking and learning, and reading avidly.

In chapter 1 I pointed out that I am not going to cover everything and I am going to rely on you to read books to further your knowledge. So now is the time to get one or both of the books by Cutmore, *Watches 1830-1980* and *The Pocket Watch Handbook*, and read them.

In addition, because there is a limit to how many watches you will be able to handle, studying photographs of watches will add to your appreciation of watch styles and help you identify the watches you do handle.

For pocket watches I suggest Reinhard Meis's *Pocket Watches From the Pendant Watch to the Tourbillon*. This book has an excellent photographic survey of pocket watches, and it also has an introductory technical section that will be useful later. Because I believe wristwatch collectors should also look at pocket watches, I think every collector should read this book.

Collecting American pocket watches is different, and it depends on whether you are collecting watches made by one of the American factories after 1850 or watches signed by individual makers before or after that time.

The history and products of the factories are well described by Donald Hoke in *The Time Museum Historical Catalogue of American Pocket Watches* and Michael Harrold's *American Watchmaking: A Technical History of the American Watch Industry, 1850-1930*. In addition, there are a number of books that give detailed information on individual companies.

In contrast, not only is there less information about individual makers, but many of the watches are not made in America. Some, like the Hamann watch, are obviously imports. However, many look like American watches but are in fact English watches with American signatures. Therefore, collecting such watches requires you to have knowledge of English as well as American watchmaking.

Unfortunately, many wristwatch books show hundreds, even thousands, of pictures of cases and dials, but they are often without dates and are sometimes presented in an almost random order. Indeed, I don't know of a wristwatch identification book that I really want to recommend. However, a good book is Kahlert, Muhe, and Brunner's, *Wristwatches: History of a Century's Development*. This book contains excellent photographs of wristwatches in a vaguely chronological order. A careful examination will enable you to recognize style trends and help date watches.

Both of these books should be browsed from beginning to end, carefully examining the illustrations. In doing so you will see trends in case and movement design, and you will also start to become aware of the dates when different designs were common.

One problem with books is that many only show high-quality, expensive watches. But as a beginner you should concentrate on ordinary, cheap watches until you have enough experience and knowledge to be able to assess watches correctly and avoid the embarrassment of spending a lot of money on a watch that is not worth it.

Homework

It is time to go out and examine some pocket watches. All you need to start with is a small pocket knife (or case opener), a loupe, and an English hallmark book that will fit in your pocket. Normally, the seller will have some keys to wind the watches that need them. Only look at simple watches without complications; leave chronographs, calendar watches, and others alone until you are familiar with watches that just display the time.

Your aim is simply to get some practice opening cases and some familiarity with watches. Find out what the case is made of, and if it has hallmarks determine the date from them. Read all inscriptions, which will usually tell you where the watch was made. Look carefully at the movement to see if it is similar to or different from other movements you have examined. And if it is similar, do the watches come from the same country? If a watch has no inscriptions to tell you where it was made, then think about the other watches you have seen and try to deduce where it came from. For example, watches with a dust cap like that in Figure 2-25, page 18, will almost always be made in England.

And don't touch wristwatches until you are confident that you can handle pocket watches carefully and without damaging them.

Finally, compare the asking prices with what you have observed, and decide if there are features that might explain the differences.

Chapter 3

The Dating Game

Signed Watches

The answers to the two questions "who made it" and "when was it made" are linked together; if we know the answer to the first question, we can usually find the answer to the second.

Dating signed watches is almost always simply a matter of looking up the signature in a book. There have been four major watchmaking countries: England, Switzerland, France, and the United States. Each country is covered in separate books, with some overlap. If you decide to specialize in watches from one country, you can limit your library to books relevant to that country. However, many collectors accidentally or deliberately collect watches from more than one country. For example, if you specialize in American watches, then nearly every watch you collect will be a simple timepiece because American watchmakers produced very few repeater watches and only a very few chronographs. So if you want to collect some watches with such complications, then you will have to include non-American pieces.

The most important books for the four regions are as follows:

United States: *American Watches Beginning to End, Identification & Price Guide,* by Roy Ehrhardt and Bill Meggers, is an excellent guide to American watches. Most American watchmaking was carried out in a relatively small number of factories, and the organization of this book is different from others. The second half of the book is a comprehensive index of signatures on movements and dials. These entries cross-reference you to the listings by company in the first half of the book, where you will find drawings of movements (*models* or *calibers*), tables of serial numbers, and dates of production.

Another book that has some value is Shugart, Engel, and Gilbert's *Complete Price Guide to Watches.* Published annually, with 25 editions to date, this book is generally regarded as the "bible" for watch prices. There is a brief introductory section with some general information, followed by sections on American pocket watches, comic and character watches, European pocket watches, and wristwatches. Because the emphasis is on prices, its value for identification is a bit limited. For example, the wristwatches illustrated are those that are "superior" and hence collectible in the eyes of the authors; for instance, there is no mention of the Unicorn brand, and often dating information is vague. But the book does include tables of serial numbers and production dates and other useful data. If you want to use this book for identification, then it doesn't really matter which edition you have, provided it is fairly recent.

In addition to these books, there are several Internet sites with information on American watches. For example, http://www.nawcc-info.org/ and http://elginwatches.org/databases/index.html have detailed serial number information for the Waltham and Elgin watches, respectively.

Remember that very few watches were made in America before 1850. Indeed, if you find an earlier watch with an American signature, it is almost certainly an English watch. If it is American, it will probably be extremely valuable.

England: The basic reference is Loomes's *Watchmakers and Clockmakers of the World 21st Century Edition.* This book amalgamates and enlarges two earlier books (with the same basic title) by Baillie and Loomes.

Although this book covers the whole world, the focus is on English makers. Each entry gives the name, address, and dates when the maker was active. Some entries have additional information.

Often there are several makers with the same surname. Under Spiegelhalter (a watch discussed in chapter 2) in volume 2 you will find ten entries of which only one fits my watch: Spiegelhalter, George (& Co.). London. 1844 (1851-1881). Usually, the correct entry is obvious, but sometimes you have to use other information, such as the style of the watch, to choose the most likely entry.

Switzerland: The best source of information is the two-volume book by Kathleen Pritchard, *Swiss Timepiece Makers 1775-1975.* It lists Swiss makers and the signatures they used, and many entries include biographies of individuals and histories of companies.

France: Tardy's *Dictionnaire des Horlogers Francais.* Being written in French is not much of a problem because we are primarily concerned with names, places, and dates. But there is some good biographical information that may be inaccessible if you cannot read French.

French watches are not our focus here. Most significant French makers worked before about 1750 and their watches are, naturally, scarce. Major watchmaking centers also were established near each other on either side of the French/Swiss border, and the border itself moved around at different times. The French imported watches from Switzerland and the Swiss often used French ébauches. Not only that, smuggling was rife! So it is very difficult to decide in which country a watch was made, and I opt for the easy solution of saying all watches are Swiss unless there is some special reason to allocate them to France.

Hallmarks, Serial Numbers, and Inscriptions

Figure 3-1.

Although signatures and hallmarks are very reliable, they must be considered very carefully. Figure 3-1 is a very interesting movement signed "John Wood, Liverpool." It is in a silver case hallmarked 1845 and with a casemaker's mark for Richard Lucas (ca. 1832-1853). However, the movement suggests a different date. First, the only possible John Wood listed in reference books worked between 1814 and 1828, and, although reference books are not perfect, the gap of 17 years is too great. Second, this is a *transitional three-quarter plate watch*. These watches are full plate, but with the balance cock flush with the plate, they look similar to three-quarter plate watches. All other transitional three-quarter plate watches that I have seen (and there are not many of them) originate from Liverpool and are dated between 1823 and 1830. Although the 1845 date may be correct, it is possible that the case was made later and this watch does fit in with the others of its type with the earlier date.

One fascinating point is that the design of this watch is very similar to the watches made by the Pitkin brothers in Hartford, CT. Starting in 1831, the Pitkins made watches using some machinery and were almost certainly the first to do so in America. It is highly likely that they based the design of their watches on an English watch like this one, just as Aaron Dennison in 1850 based his first watch, for what was to become the Waltham Watch Company, on another type of English watch.

Serial numbers and inscriptions also need to be examined with care. For example, the Hamann watch shown in Figures 2-74 and 2-75, page 30, in chapter 2 displays the following on the case dome: "Straight line lever. Full jeweled. Hands. No. 19715. J.A. Hamann, New York." Did Hamann make 20,000 watches? I very much doubt it; indeed, he probably didn't even make one! The likely explanation for the high serial numbers on watches by individuals is that these were numbers assigned by the original manufacturer. My guess is that Hamann placed an order with a Swiss company for watches with his name inscribed on them. The Swiss company simply got the next few movements coming off the assembly line and used them. Although we cannot be certain, we can suggest that the Swiss company had made about 20,000 watches at the time Hamann placed his order. This process of selling signed watches manufactured by someone else explains some of the serial numbers on English nineteenth-century watches.

Inscriptions also need careful inspection. If you have a wristwatch with the inscription on the back "To Maud love Hymie 17-4-26," you can be pretty confident of two things. First, the watch was made in 1925 or 1926, not long before Hymie bought it. Second, its collectible value has decreased because of the personal but anonymous inscription. If, however, it read "J. F. Kennedy" and you could prove the inscription was not a fake, then you would be blissfully happy.

Although inscriptions might not add to the monetary value of a watch, they can add significant emotional value. Long before the quartz revolution, watches were expensive and were treasured possessions. Unlike clocks and many other collectibles, watches are intensely personal; any watch you hold in your hands was once someone's constant companion and friend, through good times and bad. Unfortunately, by the time the collector gets a watch, all its history has usually been lost, and he or she is left with a rather anonymous example of the watchmaking craft. So the few watches that still have some of their history are to be treasured all the more.

One interesting inscription is shown in Figure 3-2. Sir John Franklin was a very famous Arctic explorer, who, although this is less well known, also lived some time in Australia as the governor of Tasmania. Surely an important inscription, right? Well, a previous owner obviously thought that it would be a good selling point, so he very crudely erased the date. In fact, the full inscription is "John Franklin, Derwent Lighthouse, Hobart 17-12-94," a date many years too late to refer to Sir John.

But all is not lost! Despite having just declined many hundreds of dollars in value, the watch is still quite interesting. My John Franklin has nothing at all to do with Sir John. He was a humble lighthouse keeper who worked on several lighthouses in Tasmania, including the Derwent lighthouse near Hobart. John worked in the north of the state from 1891 to the beginning of 1894 and

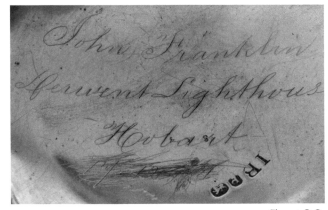

Figure 3-2.

then moved to Hobart after a two-week holiday. He either bought this watch himself or it was given to him in December 1894. This shows that government records and history books can sometimes produce a lot of useful information.

There is one other interesting feature of this watch: It was made in 1819—75 years before Franklin wore it and long before he was born! (There is no doubt about the date, which comes from hallmarks, the signature, and the style of the movement.) One suspects our poorly paid keeper bought himself the best secondhand watch he could afford. So don't assume an inscription will always date a watch correctly.

Figure 3-3.

Figure 3-4.

Another example of an inscription is that shown in Figures *3-3* and *3-4*. Although the case is badly worn, the monogram "TM" is clear and matches the inscription inside the hunter cover "Tertius Mortyn, Rodbourne, Brighton, 18-3-92." This watch and the previous watch clearly date from the nineteenth century, so there is no ambiguity about the year they were made. Because it is an English watch and Brighton is a large town in England, I initially ignored the inscription. But then I decided it might be local because there is a very small town in Tasmania that is also called Brighton. Fortunately, the surname Mortyn is unusual, and it didn't take much effort to locate his relatives.

Tertius Mortyn's father was born in 1839. He married Eliza Blacklow in 1861 and died in 1910. He was a justice of the peace and a dairy farmer at Rodbourne near Brighton; the house is still there. They had 12 children: 7 daughters and 5 sons (Lewis, Tertius, Albert, Graham, and George). Tertius Mortyn was born on March 20, 1871, and died on June 18, 1939. He fought in the Boer War and World War I and never married. He took over the Rodbourne property after his father died, and the farm remained in the family for about 100 years. From the wear on the case, we can be sure Tertius carried his watch everywhere, and probably it went with him to both wars.

But the watch and its inscriptions are mysterious. First, the watch appears to have been a twenty-first birthday present, but two days early! Perhaps the records are incorrect and he was actually born on March 18. Second, according to the case hallmarks (Figure *3-5*) the watch was made in 1849, 22 years before Tertius was born and when his father was only 10. None of the known dates for his family members match when the watch was made, but it may have belonged to a relative of his mother because the initials "AB" are crudely scratched on the inside of the back cover. But, like John Franklin's watch, it may have been bought secondhand.

This watch has a verge escapement; note the wide, flat balance rim that is commonly found on verge escapement watches (Figure *3-6*). The watch is also very ordinary, but the inscription adds somewhat to its value by making it more interesting than it would be otherwise.

Although you may not be interested in John Franklin or Tertius Mortyn, similar inscriptions more relevant to you may come your way, and the effort involved in searching records and tracing relatives can be rewarding.

Figure 3-5.

Figure 3-6.

Anonymous Watches: When Was It Made?

Dating an anonymous watch (one that is not signed or whose signature does not appear in the reference books) requires recognizing features of the case and movement that are known to have been used only at certain times.

America

American watches are easy because they are almost never anonymous. With very few exceptions they bear the name or signature of one of the watch factories and a serial number and can be dated by using the tables of serial numbers and dates that appear in many books. Most of the remaining anonymous watches were imported from England and Switzerland and are dated like any English or Swiss watch. There are some other watches that look like American watches but are in fact Swiss copies; American reference books, such as Shugart's *Complete Price Guide to Watches*, often include some information on these.

England

Most English anonymous watches are easy to date by case hallmarks if you are confident that the case is original, and using a hallmark book is essential. Philip Priestley's *Watch Case Makers of England 1720-1920* is an excellent guide because he lists casemakers working in England. Like Baillie's and Loomes's reference books, the entries give names, addresses, and dates. To use it you first need to examine the hallmarks on the watchcase and then look up the casemaker's initials. Priestley includes tables of hallmarks and instructions on how to interpret hallmarks and casemaker initials and important information about specialist casemakers.

Priestley's book can also be very useful if you come across a base metal English case. Although fairly uncommon, some watches were cased in gilt brass cases and consequently do not have hallmarks, which were used only on solid gold and silver cases. But sometimes the casemaker stamped his initials on the case, and it can be dated from information in *Watch Case Makers of England 1720-1920*.

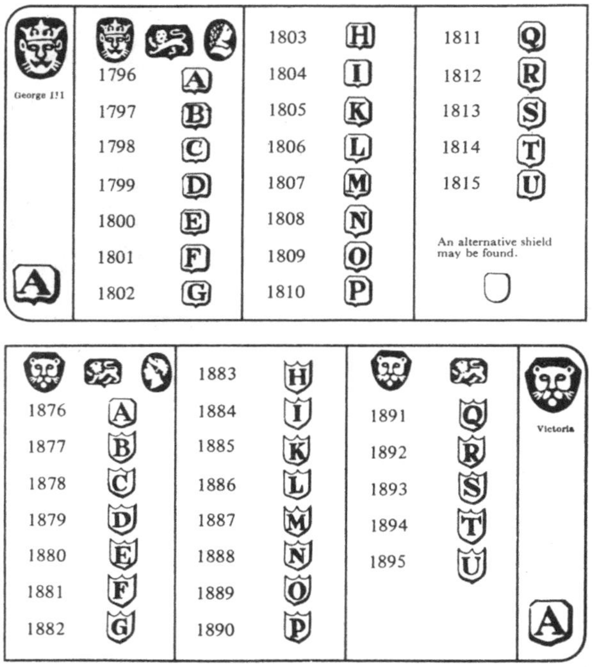

Figure 3-7.

Although dating cases is usually easy, it must be done with care. Figure 3-7 is an extract from a book of hallmarks showing the London hallmarks for 1796-1815 and 1876-1895. A quick glance at a case with a hallmark letter "C" may lead you to believe it was made in 1798 when actually it was made in 1878—some 80 years later! Two obvious features distinguish the two dates. In 1798 the leopard's head has a crown and it does not in 1878. The shape of the shield around the date letter also is different. Picking the right date can be difficult when the hallmarks are badly stamped and some variations of date shields occur, but errors can be avoided by cross-checking your choice with other information, such as the casemaker's mark and the watch style. However, some Internet sellers may deliberately misinterpret hallmarks to backdate a watch to imply that it is more valuable than it really is.

Switzerland

In contrast, Swiss pocket watches and wristwatches are often unsigned, and dating them can be difficult. Sometimes they are identified, but like the Unicorn watch we examined earlier, you may have to take the watch apart to find the signature. Even so, it is worth very carefully looking at the movement to see if part or all of a trademark or caliber number is visible. Unfortunately, the Swiss have never been concerned with collecting, except for Swatch and other modern watches. Modern "limited editions" are a popular sport of manufacturers who attempt to sell to the collectors' market. Usually, they are produced in such large numbers (20,000 or so) that they are in no way limited and are rarely worth more than the new price and often less.

Although there is a lot of practical information for repairers, it is of little use to the collector. For example, the *Official Catalogue of Swiss Watch Repair Parts* (the two-volume 1949 edition is the best) provides a huge amount of information for identifying Swiss wristwatch movements. Unfortunately, in most cases you have to take off the dial and look at the winding and setting mechanism! So you are a bit stuck because most dealers are naturally not in favor of a stranger pulling their watches to bits.

But all is not lost ...

The Dating Game

The dating game is simply an extension of what you are already doing—looking at every watch you can lay your hands on. In fact, you may have already started playing this version.

If you have followed my instructions, then you will already be aware of styles and trends. You pick up a watch, open the back, and you might immediately say "English" or "Swiss" because of the dome, hallmarks, and other features. You may also have looked at watches that are nothing at all like the ones I have described; most likely they are the rather common but fascinating *verge fusee* (named after the *escapement* it uses) with a beautifully pierced and decorated balance cock; three of these are shown below.

The dating game teaches you to order these styles chronologically. It is very simple: Whenever you look at a watch that can be dated reasonably accurately (by hallmarks, inscriptions, or serial numbers), file away in your mind, or on paper, what the movement and case look like. Now, when you next look at a watch, compare it with your memory file and estimate its date from the features it has. After you have done this, look up hallmarks, signatures, and so on to check your answer. If the date agrees with your estimate, give yourself a treat! If you are wrong, reassess the watch and its design and file away the new information you have.

For example, the Lemania Chronometer shown in chapter 2 as Figures 2-37 to 2-40, page 21, is from World War II, circa 1940. Compare it to the Hamilton watch shown in chapter 2 as Figures 2-41 to 2-44, page 22, which was made in 1912. There are many differences, but I note two that are important.

The pivot jewels in the Hamilton watch are mounted in small disks of metal, called chatons, and these are held in place by small screws. In contrast, the pivot jewels in the Lemania are set straight into the top plate. Indeed, it is a general rule that watches using chatons are earlier than watches with pressed-in jewels.

However, this feature must be treated with care and what applies to one country may not apply to another. If you look at only American pocket watches, you will see a definite trend from chatons to pressed-in jewels. You will also see a definite trend in plate engraving; early watches have plain, gilt brass plates. Then highly ornamental damaskeened nickel plates appear. And later the plates (and the jeweling) revert to being fairly plain. But if you look at a Swiss watch with pressed-in jewels, don't apply the same rule and decide it is from the middle of the twentieth century. It may well date from the 1880s.

foot

Figure 3-8.

The second point is that full-plate watches are generally earlier than three-quarter or half-plate watches, with the same restriction that you must be comparing like with like.

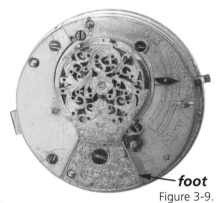

foot

Figure 3-9.

England

The shape and style of balance cocks are important tools for dating and identifying early watches. In Figures 3-8 and 3-9 two verge movements are shown. The Figure 3-8 watch shows the foot of the balance cock as very wide and pierced, which is typical of the early eighteenth century (1700-1750); this watch is ca. 1740. Figure 3-9 has a narrower, unpierced foot, typical of the late eighteenth century, with some persisting into the early nineteenth century. Note that the cock completely covers the balance, whereas later cocks leave much of the balance exposed. After 1800 nearly all verge watches have a narrow, simple cock like those we have seen on other watches; the watch in Figure 3-6, page 37, is a good example. Figure 3-10 highlights the *contrate wheel* with its teeth standing up vertically to the wheel. Whenever you see a contrate wheel, it is almost certain that the watch uses a verge escapement. *Almost* is the operative word. There are some rare watches with contrate wheels that use other related escapements.

contrate wheel

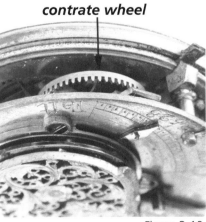

Figure 3-10.

Figure 3-11.

Figure 3-12.

Figure 3-13.

Figure 3-11 shows another verge watch that was made around 1770. This time the balance is supported by a bridge with a foot and screw at each end, but the screw for the left foot is missing. Just to confuse you, such balance bridges are often called *balance cocks* even though they are not cocks, and some people call balance cocks *bridges*, even when they have only one foot! However, there is a very useful rule: If there is a balance bridge with two feet, then the watch is almost certainly a *continental watch* made in Europe (usually Switzerland or France); if there is a cock with one foot, the watch was almost certainly made in England.

This continental watch is signed "Wilders, London." It comes from a time when many watches made on the continent were signed with false English names and sold in England. There are also watches like this signed Tarts and Farts, although I suspect they would not fool an English speaker.

The dial of the watch in Figure 3-12, which is badly chipped at the numeral I, is also interesting. It is enamel and the scene on it was painted by using different colored crushed glass and a very fine paint brush. Then the dial was fired in a kiln to melt the painting onto the white glass background. Even crude paintings like this one required a great deal of skill. The wavy pattern of fine lines at the edge of the dial and running inside the minute numerals is also interesting. This style of chapter ring is called *arcaded minutes* and was used in Holland for a fairly short period of time. It confirms the continental origin of the watch.

Other examples of balance cocks are shown in Figures 3-13 to 3-17 where five English watch movements are shown. All are full plate of varying quality (some jeweled and some not), but all have differently shaped balance cocks, which show a stylistic trend. They all use some form of *lever escapement*. Figure 3-13 has an ornately shaped and pierced cock and was made in 1798. Figure 3-14 has a similar cock (but not pierced) and was made in 1817. Later, in Figure 3-15 (this example is ca. 1840) the cock is still shaped like a wedge, but it has become narrower. And finally, in Figure 3-16 in the late nineteenth century the cock became narrower still, more uniform, and less highly ornamented or even plain. Figure 3-17 is an example of a number of watches that were made with a straight cock around the 1840s.

These movements have a number of other interesting features, but I will only comment on three. First, Figures 3-14, 3-15, and 3-17 have very large, obvious pivot jewels in chatons. These are called *Liverpool windows* and can also help to date a watch. Second, I said above that the movements are "of varying quality (some jeweled and some not)." In fact, Figure 3-13

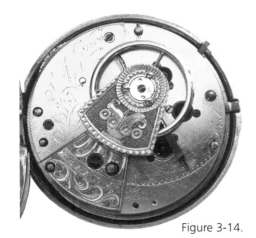

Figure 3-14.

Figure 3-15.

is probably the most important and most interesting of these movements, but only the balance cock has a jewel. To judge quality by jewels alone can be very misleading, and we shall look at this later. This watch has a "thing" under the balance, which is a screw and a slot as shown in Figure 3-18. This feature is found on watches with a *rack lever escapement.* Third, two of the watches (Figures 3-16 and 3-17) display the telltale banking pins, which indicate a lever escapement.

Later English watches generally have three-quarter plate movements, like those in Figures 3-19 and 3-20, made in 1888 and 1900, respectively. The left movement is made by J. W. Benson "By warrant to H.M. the Queen." It looks as though it has a fusee because of the extra pivot marked with an arrow, but this is a *dummy fusee* watch with an extra wheel to reverse the direction of winding. Many people assume a watch with the winding square on a separate arbor to the barrel has a fusee, and I have often seen Benson "Ludgate" watches like this one incorrectly described as fusee watches. The right movement, made by William Ehrhardt, is pendant wound and does not have a fusee.

Except for one being key wound and the other pendant wound, the movements are almost mirror images of each other. Under each balance there is a curved cock that supports the pivots of the lever and escape wheel.

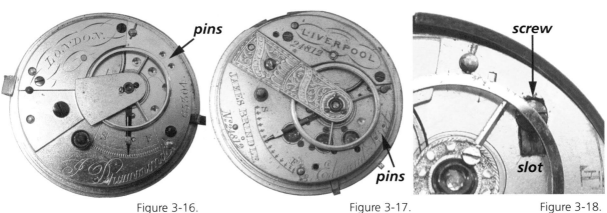

Figure 3-16.

Figure 3-17.

Figure 3-18.

Figure 3-19.

Figure 3-20.

In addition to Meis's *Pocket Watches From the Pendant Watch to the Tourbillon*, there are four books on English watches that are well worth reading:

If you are like most collectors, you will begin by acquiring watches that are easy to obtain and not too expensive, and these will mainly be watches made after about 1850 (for pocket watches) or 1940 (for wristwatches). An excellent book covering this period is M. Cutmore's *Watches 1830 to 1980,* which is a very well-written, well-organized history of watchmaking illustrated by the more common watches that you are likely to encounter. It is a must-read book.

Another book that concentrates on this period is Alan Shenton's *Pocket Watches 19th & 20th Century.* This very good book is organized around excellent, captioned photographs, and it is a very clear study of watch styles.

Baillie's *Watches—Their History, Decoration and Mechanism* is almost exclusively concerned with watches before 1725, so it concerns timepieces that are very hard to find and very expensive. Although I will not be looking at any watches made before 1750, this excellent text should be read by all collectors.

Finally, another essential book for your library is *Britten's Old Clocks and Watches and Their Makers* (ninth edition) by Baillie, Ilbert and Clutton. The early editions were written by Britten alone and are mediocre; they should be avoided unless you collect books. But the ninth edition, with the text completely rewritten by G.H. Baillie, C. Ilbert, and Cecil Clutton, is probably the best book written on pre-1830 watches. The seventh and eighth editions may be all right, but I haven't read them and cannot comment.

America

Dating and identifying American watches is, as I have pointed out, easy. For example, Figures 3-21 through 3-26 show six American movements dating from 1884 to 1930, and all are quite precisely dated by their serial numbers. Figure 3-21 is an ordinary, gilt brass, full-plate Waltham dated 1884. The two screws marked by the arrow are the ends of the banking pins; most full-plate American watches have banking pins that are screwed into the plate. Figure 3-22 is a damaskeened nickel full-plate movement from ca. 1890. The damaskeening and the imitation "gold" chatons make it look better than the Waltham in Figure 3-21, but in reality it is not quite as good. Note that the chatons are pressed in and not held by screws. Figure 3-23 is a *divided half-plate* from 1905, and this time the chatons hold jewels; it is a much higher quality movement than the previous two.

Figures 3-24 and 3-25 are the same *model* or caliber, but the earlier movement in Figure 3-24 (from 1917) is of higher quality than the later movement in Figure 3-25 (from 1921); it has jewels in chatons and a different type of regulator on the balance cock (indicated by the arrows) for adjusting the balance spring. The damaskeening is nearly identical. Figure 3-26 is a smaller watch made in 1930 and has a *divided three-quarter plate*. In addition to *American Watches Beginning to End: Identification & Price*

Figure 3-21.

Figure 3-22.

Figure 3-23.

Figure 3-24.   Figure 3-25.   Figure 3-26.

Guide by Ehrhardt and Meggers, there are two excellent books that every collector should read to develop an understanding of the progressive changes in American watches:

D. R. Hoke's *The Time Museum Historical Catalogue of American Pocket Watches* begins with two long essays on the American watch industry and then catalogs the Time Museum collection (which unfortunately has been split up). The superb color plates provide an excellent survey of American watchmaking to complement Ehrhardt & Meggers's *American Watches Beginning to End, Identification & Price Guide.*

Michael Harrold's *American Watchmaking, A Technical History of the American Watch Industry 1850-1930* is an excellent, detailed history of American makers.

Switzerland

The Swiss watches below date from circa 1780 to circa 1965. Figure 3-27 is a typical eighteenth-century full-plate verge with the balance cock having two screws. Such movements often have a small cock **d** on the side. This is the *counter potence,* which supports one end of the escape wheel arbor. Also note that this watch has a steel plate **e**, the *coqueret,* on the balance cock instead of a jewel. Figure 3-28 is a three-quarter plate movement from 1870 cut out to look as though it has three cocks. Swiss watches are rarely damaskeened and most have plain or engraved plates and cocks. Figure 3-29 is a typical barred movement, circa 1890.

Figure 3-27.   Figure 3-28.   Figure 3-29.

Figure 3-30.

Figure 3-31.

Figure 3-32.

Figures *3-30* and *3-31* show two divided three-quarter plate movements. The watch shown in Figure *3-30* is circa 1910 and that in Figure *3-31* is circa 1965. Note the very bad marks on the left, near the pendant, of Figure *3-31*. The snap on the inner, hinged dome is very tight and carelessness has caused quite deep gouges. Also note the pressed-in jewels. Figure *3-32* is the dial of the watch in Figure *3-31*. It is signed "Cortebert T.C.D. Demiryolu" and the back of the case has "T.C.D. Demiryollari Anti-magnetic Swiss Made" with the Turkish crescent and star surrounding an engraving of a train. These watches were made for the Turkish national railway "Turkiye Cumhuriyeti Devlet Demiryollari," but this one was used by the Western Australian Government Railway.

The last watch, Figures *3-33* and *3-34*, is another divided three-quarter plate movement, circa 1940. It is almost identical in layout to the Waltham watch in Figure *3-25*, page 43.

Figure 3-33.

Figure 3-34.

The three watches in Figure *3-30* to *3-34* span some 50 years and are almost identical in design. There is almost no way that such watches can be dated accurately by style alone. Figure *3-34* shows the case of the movement in Figure *3-33*, which appears to be very similar to the Waterbury watchcase shown in chapter 2 in Figures *2-31* and *2-32*, page 20. However, despite the lack of anything to indicate it, this case has a screw back and bezel! It is not surprising that there are some marks from attempts to prize the covers off.

The best sources of information on Swiss watchmaking I know of have already been mentioned: Meis's *Pocket Watches From the Pendant Watch to the Tourbillon*, Kahlert, Muhe, and Brunner's *Wristwatches, History of a Century's Development*, Cutmore's *Watches 1830 to 1980*, and Shenton's *Pocket Watches 19th & 20th Century*.

Wristwatches

A part from modern watches, mechanical wristwatches existed for some 60 years, from about 1915 to 1975. Even so, dating them can be more difficult than dating pocket watches.

Most of the names on pocket watches are of individual people or companies, and it is fairly easy to find out when they lived and worked. This information, together with other indications, such as style, serial numbers, and hallmarks, often helps us date a watch within a few years. Most of the companies that manufactured watches over a long period of time also kept information about serial numbers and production dates, as with the Waltham Hillside we examined in chapter 2, Figure 2-7, page 13.

Most wristwatches were manufactured by large companies in Switzerland and Japan, with smaller numbers from America, England, France, Germany, and Russia. Although some production information may be available, often we cannot open the case to inspect the movement and find the serial number and other information we need. And there are vast numbers of wristwatches, like Figure 2-45, page 23, shown in chapter 2, that appear to be completely anonymous (actually this watch has the ébauche maker's trademark under the dial). To a much greater extent, wristwatch dating is based on appearance and styles. Because they are a fashion item, styles changed regularly to suit the tastes of people, and these styles can often limit the date of a watch to just a few years.

There are some trends in design that allow you to decide one watch was made before another. For example, hinged cases were only used during the transition from pocket watches, screw backs were introduced much later, and fully sealed cases (single-piece cases where the crystal has to be taken off to get at the movement) are later still.

However, many styles have been resurrected to suit changes in fashion and the production of retrospective series. Consequently, it is extremely difficult to date some wristwatches by appearance, and unless you have details of a company's production or serial numbers, it is often hard to be sure of the age of a particular watch. For example, if you look carefully at the illustrations of dials in Kahlert, Muhe, and Brunner, *Wristwatches, History of a Century's Development*, you will see some virtually identical watches dating from 1920 and 1950. Opening the case and examining the movement may enable you to fix a date, but often this is impossible to do.

As a result, dating wristwatches is very much a matter of experience, of having looked at large numbers of dated watches so that you gain an intuitive feel for their ages.

Chapter 4

The Movement Game

Trains

The movement game is an extension of what has been discussed so far in this series. So far you have looked at watches to study styles and dating. Now we shall examine the movement to determine its features. If you have access to the Internet, this game can be played with great success on eBay. You will have the additional pleasure of discovering many watches for sale that are incorrectly dated and described, and you will be able to determine the country of origin by style rather than by the signature.

Before we look at trains, let me note that most of the photographs in this section are of a scrap movement (Figure 4-1), which will soon be consigned to the rubbish bin. It is a typical Swiss barred movement of reasonable quality from the 1880s, with a cylinder escapement and ten jewels. At first glance it is anonymous, but there is a trademark on the pillar plate (marked by the arrow) that includes the letters "AW" (Figure 4-2). My first guess was that this would be the A... Watch Co., so I looked it up in Pritchard's *Swiss Timepiece Makers 1775-1975,* but found nothing.

I then tried another book, Kochmann's *Clock and Watch Trademark Index - European Origin.* This book is of more use for identifying clocks than watches, but it can still be very helpful. Its index listed two "AW" marks: one for Germany, one for England, and none for Switzerland! I checked both, just in case, and the English entry was the right one. The signature is for Adolphe Woog, London, and the trademark was registered in 1876. Woog is described as an importer and also had an office in Paris. Despite his address this is definitely not an English watch.

The moral of this story is: Just because you are now looking at movements, do not neglect style, identification, and dating.

Watch movements contain three basic features: a *train* of wheels and pinions, a source of power (the *mainspring*), and a mechanism to control the rate of movement of the train (an *escapement*).

Figure 4-2.

inset (Figure 4-2)

Figure 4-1.

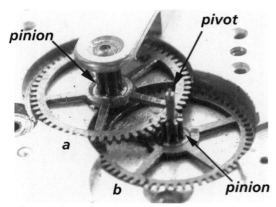

Figure 4-3.

Figure 4-4.

The principle of a train is simple. In Figure 4-3 we can see two brass wheels with teeth around their edges. These wheels are mounted on *arbors,* and at each end of the arbors are *pivots,* which fit in holes in the plates or cocks; one cock has been removed to show the pivot. Also on the arbors are small, steel wheels called *pinions.* As a wheel rotates, its arbor and pinion rotate with it.

The teeth on wheel **a** mesh with the teeth, or *leaves,* of the pinion on the second arbor. Let us suppose wheel **a** has 50 teeth around its rim and the pinion it meshes with has 10 leaves. As **a** turns, each tooth moves the pinion around by one leaf. If 10 teeth of **a** move past the pinion, then the pinion rotates by 10 leaves, which is a full revolution. Thus, if the wheel rotates once, the full 50 teeth, then the pinion will rotate five times.

Now, there is another wheel **b** mounted on the arbor of this pinion, and this second wheel **b** will also rotate five times for every turn of wheel **a**.

In Figure 4-4 I have added another wheel **c**. Let us suppose this wheel has 60 teeth and the pinion it meshes with, on the arbor of wheel **a**, has 10 leaves. Then every time wheel **c** rotates once, the pinion it meshes with and the wheel **a** will rotate six times. As **a** rotates, it turns wheel **b**, and for six rotations of **a**, wheel **b** will rotate 30 times. So, for each turn of wheel **c**, wheel **b** rotates 30 times. These wheels collectively form a train.

Now suppose wheel **c** rotates once in an hour. Then wheel **b** will rotate 30 times in an hour, or once every two minutes. If we change the number of teeth and leaves on the wheels and pinions, we can get wheel **b** to rotate any number of times in an hour that we want. For example, if we had the same wheels but the pinion on the arbor of **b** only had five leaves, then wheel **b** would rotate 60 times in an hour, once every minute.

Also, a single wheel and pinion also would do. If wheel **c** had 600 teeth and the pinion on arbor **a** had 10 leaves, then wheel **a** would rotate 60 times for each rotation of **c**. But wheel **c** would be huge! The purpose of using a *train* of wheels and pinions instead of a single wheel and pinion is simply to be able to use wheels and pinions of convenient sizes.

In addition to wheel **b** rotating faster than wheel **c**, the amount of power at wheel **b** is proportionately less. A crude analogy is to note that the total power applied to turn **c** once is distributed over 60 rotations of **b**, and each rotation of **b** involves one sixtieth of the energy.

Horology is the science of time measurement. But in fact horology is primarily practical mechanics with very little useful theory. Sometimes a theory, such as for the shape of wheel teeth, is interesting and useful, but most watchmakers were unable to apply it to the tiny objects they created, and so they used approximations, which were often crude. This is why you will come across statements saying teeth should be the shape of bay leaves or thumbs! On other occasions a theory, such as for the behavior of oscillators, is so abstract it has only borderline relevance to a real balance and spring.

However, there is one bit of theory that you will find discussed in detail in many books: the theory of gear trains. To design a gear train you need to know nothing except how to multiply and divide—things that most people can do. Because it is simple, many books include it, not because it is particularly useful, but because it gives an air of technicality to books that are otherwise purely descriptive!

The following two books describe basic train calculations and are readily available:

Anthony Whiten, *Repairing Old Clocks and Watches.* This is a book written by an amateur for amateur collectors. Regardless, it is one of my recommended books, because Whiten is very good at describing basic repair tasks.

Donald de Carle, *With the Watchmaker at the Bench.* This is a book written by a professional for teachers. It is worth reading in the future, but not just yet. But if you have access to a copy, read the section on trains.

The Barrel

Figure 4-5 shows a complete watch train. To the other three wheels has been added the mainspring *barrel.* This barrel is a *going barrel* and has teeth around its rim so that it acts like a wheel that meshes with the pinion on arbor **c** in Figure 4-4, previous page. The wheel on the barrel is called the *great* wheel or the *first* wheel. Wheel **c** is the *center* or *second* wheel; it is almost always in the center of the movement. Wheel **a** is the *third* wheel and wheel **b** is the *fourth* wheel. Still missing from this watch are the balance and escape wheel, which make up the regulator to control the speed at which the train rotates.

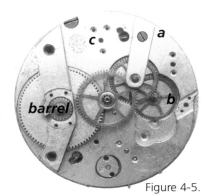

Figure 4-5.

Figure 4-6 shows the *mainspring* coiled up inside the barrel; the barrel has a lid that has been removed for this photograph. The inner end of the mainspring is hooked onto the barrel arbor, and this arbor has a square on the end. The outer end of the mainspring is hooked onto the wall of the barrel. If I hold the barrel still and rotate the barrel arbor clockwise, the mainspring will tighten until it is wrapped around the arbor. If I then let go of the barrel, the tension in the mainspring will force the barrel to rotate clockwise until the mainspring is again relaxed against the barrel wall. Of course, I have to hold the barrel arbor stationary or it too will rotate rapidly.

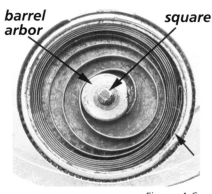

Figure 4-6.

With the barrel in the movement, the mainspring will cause the barrel to rotate the wheels of the train. Because there is no way of controlling what happens, the wheels will rotate at high speed. If the watch has hands on it, they will whiz around at an alarming rate!

For the moment, let us assume there is some sort of controlling mechanism and look at this train.

Watches usually run for about 30 hours on a single winding of the mainspring. Because the barrel is quite small, the mainspring can only have a few turns, which provide the power for over a day's running. Suppose it has five turns. Then each turn of the barrel must take six hours so that five turns will last 30 hours. Now, the center wheel has the minute hand attached to it and must rotate once every hour. Consequently, for each turn of the barrel, the center wheel must rotate six times, and the number of teeth on the barrel and the center wheel's pinion are chosen to achieve this.

If the watch has a seconds hand, it is normally attached to the fourth-wheel arbor. Because the seconds hand rotates once a minute, the fourth wheel needs to rotate 60 times for one rotation of the center wheel. The numbers of teeth on the wheels and leaves on the pinions are chosen so that this happens.

If the watch doesn't have a seconds hand, then it doesn't matter how fast the fourth wheel rotates.

As I have mentioned, the barrel will only rotate if the barrel arbor is prevented from rotating. This is achieved by using a *ratchet* and *click*.

Figure 4-7.

Figure 4-7 is a closeup view of the barrel bridge. Fixed to the barrel arbor is a *ratchet wheel,* and a *click* acting in the teeth of this wheel prevents it from rotating counterclockwise. When the watch is wound by a key on the winding square, the barrel arbor can be turned clockwise with the click sliding over the teeth of the ratchet, but when you stop winding, the arbor cannot turn the other way and let down the mainspring. So the only way the tension in the mainspring can be released is by rotating the barrel and the train. This click and the spring that holds it against the teeth of the ratchet is made from a single piece of steel; however, clicks and their springs come in many different shapes and sizes.

Note the recess around the ratchet and the three holes. In fact, there was a cover plate held by three screws over the ratchet wheel, which I have removed.

Watches with *fusees* do not have a going barrel, and the great or first wheel is attached to the bottom of the fusee (Figure 4-8). The barrel arbor is stopped from turning by a ratchet, click, and spring (Figure 4-9). In most watches these are placed under the dial and are not visible, but for a while they were placed on the top plate as a decorative feature.

Figure 4-8.

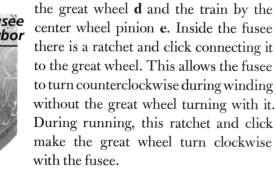

Figure 4-9.

The fusee chain, **b** in Figure 4-8, is wrapped around the barrel and its end hooked into a hole below **c** in the side of the barrel (the semicircular hole in the barrel lid is put there to enable the lid to be prized off). When the watch is wound, by turning the square **a** on the fusee counterclockwise, the chain **b** connecting the fusee to the barrel is drawn off the barrel onto the fusee, winding the mainspring. When the watch runs, the tension of the mainspring pulls the chain off the fusee, turning the great wheel **d** and the train by the center wheel pinion **e**. Inside the fusee there is a ratchet and click connecting it to the great wheel. This allows the fusee to turn counterclockwise during winding without the great wheel turning with it. During running, this ratchet and click make the great wheel turn clockwise with the fusee.

Motion Work and Keyless Work

The center wheel of the watch I have described in Figures 4-5 to 4-7, above, rotates once each hour, and all the other wheels of the train rotate faster than this. So we can have a minute hand, but as yet there is no way to have an hour hand that rotates once in 12 hours.

Figure 4-10.

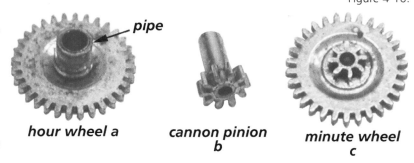

hour wheel a   *cannon pinion b*   *minute wheel c*

We could attach a minute hand to the center wheel arbor. But if we did, there would be no way of setting that hand to the right position to align it with the markings on the dial. To turn it, we would have to turn the whole train, including the barrel, which is not possible.

To overcome this, a special train, the *motion work*, is used. The motion work consists of three components (Figure 4-10). The cannon pinion **b** is a hollow tube that fits firmly over the center wheel arbor. This arbor, shown at Figure 4-11**d**, extends through the bottom plate, and it is long enough to protrude above the dial. The friction between the cannon pinion and the center wheel arbor means that the cannon pinion will rotate with the center wheel and a hand on it will rotate once each hour. But the friction between them is not great enough to prevent the cannon pinion being turned around while the center arbor remains stationary, so that the hands can be set to the right time.

Figure 4-11.

Two wheels are also used. The *minute wheel* **c** sits beside the cannon pinion **b** and meshes with it (Figure 4-12). The *hour wheel* **a** sits loosely over the cannon pinion and meshes with the pinion on the minute wheel (Figure 4-13). Finally, the hour hand fits on the *pipe* of the hour wheel and the minute hand goes over it, attached to the cannon pinion (Figure 4-14).

The minute and hour hands are geared together through this motion work, and if one is rotated, the other must also turn. But because the cannon pinion can turn around on the center wheel arbor, the hands can be moved and set to time without affecting the watch train.

If you count the number of teeth and leaves on the three components in Figure 4-10, you see the following: The cannon pinion has 10 leaves and the minute wheel 30 teeth, so for one revolution of the cannon pinion the minute wheel turns one-third of a revolution. The pinion on the minute wheel has 8 leaves and the hour wheel has 32 teeth, so one turn of the minute wheel will rotate the hour wheel one-quarter of a turn. And so one turn of the cannon pinion will rotate the minute wheel one-third of a turn, which will rotate the hour wheel one-twelfth of a turn. Which is just what we need for the hour hand.

Figure 4-12.   Figure 4-13.   Figure 4-14.

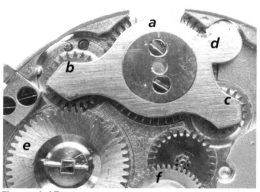

Figure 4-15.

Figure 4-16.

The *keyless mechanism* consists of the crown, stem, levers, and wheels, that enable a watch to be wound and set without a key. There are two basic systems, the *rocking bar* and the *shifting sleeve*.

In Figures 4-15 and 4-16 the rocking bar is a steel plate with three wheels **a**, **b**, and **c** attached to it, and it is pivoted in the center, which is also the center of **a**. The three wheels **a**, **b**, and **c** are always in mesh, and a spring **j** keeps the rocking bar rotated counterclockwise so that wheel **b** is in mesh with wheel **e** on a square on the barrel arbor. The crown and stem turn wheel **a** through an intermediate wheel **g** placed at right angles to **a**. On the outside of the case there is a small push-piece that can be pressed in by a fingernail. It acts on **d** to swivel the rocking bar so that **c** meshes with the motion work

Figure 4-17.

stem

Figure 4-18.

f. As soon as the pressure on **d** is released, the rocking bar moves back to the position shown in Figures 4-15 and 4-16.

Through the gears on the rocking bar, the crown is normally in mesh with the barrel arbor and winds the watch. Because the barrel arbor only rotates during winding, the rocking bar can stay in this position without affecting anything.

In contrast, the motion work wheels are always in motion. If the rocking bar, and hence the crown, remained in mesh with **f**, the crown would rotate all the time the watch was running. The friction of the keyless mechanism and of the crown against clothing would cause the cannon pinion to slip on the center wheel arbor or the watch to stop.

Figure 4-15 is the keyless mechanism from a good watch (made by Longines). Figure 4-16 is the mechanism of a cheap, badly made watch. Note that flat bits of steel are used for the spring **j** and the extremely crude click **h**.

Figures 4-17 and 4-18 show a *shifting sleeve* keyless mechanism that is lever set. The stem attached to the crown has a square on it, which fits inside the *castle pinion* **a**. In the normal winding position, the castle pinion is held up against the wheel **e** by the lever **d** and the spring **b**; wheel **e** connects to the barrel arbor. When the crown is turned, **a** rotates and the mainspring is wound. The meshing teeth of **a** and **e** are ratchet shaped so that if the crown is turned the wrong way, they slip over each other.

When the setting lever **c** is pulled out, it forces the lever **d** down and the castle pinion goes out of mesh with **e** and into mesh with the motion work.

There are a large number of different arrangements for these two keyless mechanisms, using the crown, pushpins, and set levers to activate them. Good descriptions of the basic variations are found in Britten's *Watch and Clockmakers' Handbook, Dictionary and Guide* and Cutmore's *Watches 1830-1980*.

Fusee watches rarely have keyless mechanisms and are nearly always wound and set with a key. Because the fusee turns both during winding and running, it is not possible to have a crown permanently attached to it, as can be done with a going barrel. Consequently, it is very difficult to devise a mechanism that works effectively and allows the necessary freedom.

Automatic or self-winding watches are almost exclusively the province of wristwatches. Self-winding pocket watches were developed at the end of the eighteenth century, but they are very rare. Except for a couple of early designs, all wristwatches use a semicircular, pivoted weight, the rotor, which turns whenever the watch is moved and winds the mainspring through a special train of gears. Nearly all such watches look like the Bucherer watch in chapter 2, Figure 2-55, page 25, and the two watches in Figures 4-19 to 4-21 where the rotor is pivoted in the center of the movement. The movement of some early rotors is limited to less than 360 degrees by two buffers, but most swing completely freely. In addition, some systems only wind the watch when the rotor moves in one direction and do not wind it when turning the other way.

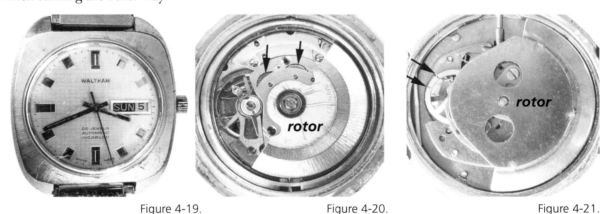

Figure 4-19.   Figure 4-20.   Figure 4-21.

Figures 4-19 and 4-20 are the dial and movement of a Swiss-made Waltham watch; after the American Waltham company closed in 1957, the name was purchased and used on Swiss watches. This is a good-quality watch with 25 jewels. The two winding wheels, which are turned by the rotor, are marked by arrows in Figure 4-20. Figure 4-21 is the movement of a Timex watch that probably has no jewels. The movement plates and rotor are made of thin steel. The two arrows point to small holes drilled part way through the rim of the balance, which I will explain shortly.

Although superficially similar, there is considerable variation in the details of self-winding mechanisms, and books that describe them and their repair have been written. Although they are mentioned in Kahlert, Muhe & Brunner's *Wristwatches: History of a Century's Development*, I do not know of a suitable introductory book. But there is an excellent book for later on: *The History of the Self-Winding Watch 1770-1931*, by Jaquet & Chapuis, is a superb study and should be read by everyone.

The Balance and Balance Spring

As I have already pointed out, after the mainspring is wound, the train will run down very rapidly unless there is some way of controlling the speed. The controlling mechanism in all mechanical watches after about 1675 consists of a balance with a balance spring (also called a *hairspring*) and an *escapement*.

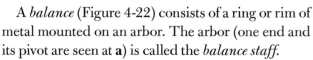
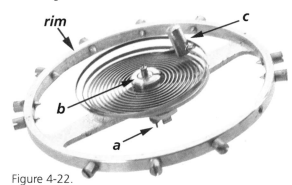

rim

c

b

a

Figure 4-22.

A *balance* (Figure 4-22) consists of a ring or rim of metal mounted on an arbor. The arbor (one end and its pivot are seen at **a**) is called the *balance staff*.

The *balance spring* is usually a spiral spring. The inner end is attached to the *collet* **b**, which fits tightly on the balance staff. The outer end of the spring is attached to the *stud* **c**, which is fixed to the balance cock or the top plate.

With the *balance complete* (balance, balance staff, and balance spring) in the movement, it will sit without moving in its at-rest position. If we turn the balance a little to one side or the other, the spring will become tensioned, and when we let go, it will force the balance to rotate. The momentum of the balance will cause it to swing past the rest point until the increasing tension in the spring brings it to a stop, when the spring causes the balance to swing back the way it had come. How long it takes to swing from side to side, the frequency, depends on the weight of the balance and the strength of the balance spring.

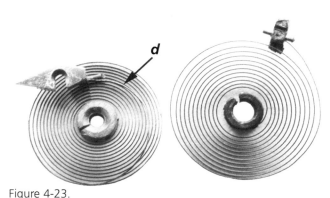

d

Figure 4-23.

The oscillations of the balance control the rate at which the train runs and so the time shown by the hands. In an ideal world every oscillation would take exactly the same amount of time and the watch would keep perfect time, but several factors prevent this.

Changes in *temperature* cause the elasticity of the balance spring to alter and the size of the balance to increase or decrease slightly, affecting its momentum; and so a watch will run at different rates at different temperatures. *Plain* balances (Figure 4-24**a**), which are simply a ring of brass, steel, or gold and which are found frequently in old pocket watches and cheap wristwatches, cannot compensate for temperature changes, and watches with them are not very accurate. *Compensation* balances (Figure 4-24**b**) try to overcome this. They have a rim composed of two metals, usually brass and steel, with screws around the rim. The rim is cut through

Figure 4-24.

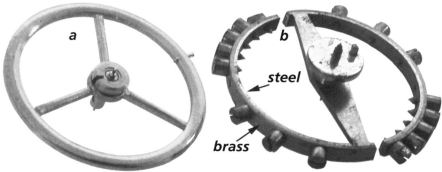

a

b

steel

brass

near the balance arms. The cut ends can move in and out with changes in temperature and reduce the variations in the time of oscillation. The screws around the rim are used to adjust the effect of this compensation. Plain balances are also made with screws (as in Figure 4-22, page 54), and bimetallic compensation balances are sometimes uncut, but neither can compensate for temperature changes. Modern *monometal* balances look like plain balances, but they, and the balance springs used with them, are made of special alloys and are almost completely unaffected by temperature. They are better than compensation balances but still not perfect.

As the *mainspring* runs down, the power available to the balance decreases, and so the size of its oscillations decreases. Ideally, every oscillation, no matter how large or small, should take exactly the same time; this is called *isochronism*. However, perfect isochronism is impossible, and watches run at different rates when fully wound to run down.

Gravity also affects the oscillations, and this changes when the watch is held in different positions. Its effect on balances is significant and they are poised to reduce it. This means that all points around the rim have the same weight and there are no heavy or light spots. If there is a heavy spot, then gravity will cause the balance to try to rotate so that it is at the bottom, and this can seriously affect the accuracy of the watch. Poising balances with screws is done by altering the weights of the screws. Balances without screws can only be poised by removing metal from the heavy spots, which is why the balance in Figure 4-21, page 53, has two small holes in the rim (see arrows). Gravity has other effects, including causing the balance spring to sag and affecting the friction between the pivots and their holes.

Balance springs are commonly *flat* spirals or overcoiled, as shown in Figure 4-23, page 54. The last turn **d** of the overcoil spring, which is also called a *Breguet* spring, is bent up and then curves in closer to the center. This is done to minimize the effects of gravity on the spring. (The balance spring attached to the balance in Figure 4-22 is actually a flat spring, but the stud got caught inside the second turn of the spring.)

The Escapement

To use the oscillations of the balance to control the watch train, so that it runs slowly and evenly and displays the time accurately, we need an *escapement*.

All escapements have two features:

1. The last wheel of the train drives the escape wheel **e** (Figure 4-25); in this example it is a cylinder escape wheel. The escape wheel is held in check, but at some point in the balance's oscillation the escapement will allow the watch train to rotate by releasing just one tooth of the escape wheel. If the balance oscillates four

Figure 4-25.

   times per second, then the escape wheel will move one tooth every quarter of a second. And we can design the wheels and pinions of the train so that this motion is converted into one rotation per hour of the center wheel.

2. Friction between the balance staff pivots and their holes, gravity, and the resistance of the air, to name just three factors, will rapidly bring the balance to rest. So the power in the mainspring, which causes the train, including the escape wheel, to rotate, is used to impulse the balance to keep it swinging.

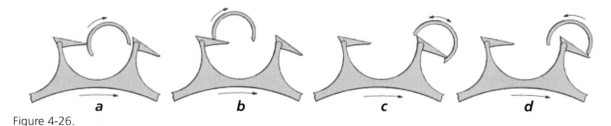

Figure 4-26.

To illustrate this, Figure 4-26 shows how the cylinder escapement works; I have chosen this escapement because it is very easy to understand.

The balance staff is a hollow cylinder, of which a little less than half has been cut away, and the escape wheel has triangular teeth. In Figure 4-26**a**, the balance has rotated as far as it can go counterclockwise and has started returning clockwise. The escape wheel is trying to rotate clockwise, but its movement is blocked by the tooth resting on the outside of the cylinder. Eventually, it rotates enough so that the tip of the tooth slides past the lip at the end of the cylinder, Figure 4-26**b**. Now the escape wheel can rotate, and the power of the train causes the tooth to push against the lip and force the balance clockwise, providing an impulse. However, when the balance has rotated far enough for the escape wheel tooth to slide past, Figure 4-26**c**, the tip of the tooth lands on the inside of the cylinder and the escape wheel is again blocked. The balance continues rotating clockwise until the tension in the balance spring stops it and makes it return counterclockwise. Again it rotates enough so that the tip of the tooth inside the cylinder slides past the lip and impulses the balance counterclockwise, Figure 4-26**d**. When the balance has rotated far enough for the escape wheel tooth to slide past, the tip of the next tooth lands on the outside of the cylinder, and we are back at Figure 4-26**a**.

Escapement Types

The most important thing to look at is the escapement, because the escapement is used to classify watches; for example, watches are often described as English levers or Swiss cylinders.

The most common escapements are those shown in Figure 4-27 and Figure 4-28.

Figure 4-27 shows the escape wheel and lever of three basic variants of the lever escapement:

The *English* lever **a** has pointed teeth on the escape wheel, and the two pallets are in line with the body of the lever. Although the balance is not shown, a line drawn from the escape wheel arbor to the lever arbor is at right angles to a line from the lever arbor to the balance staff.

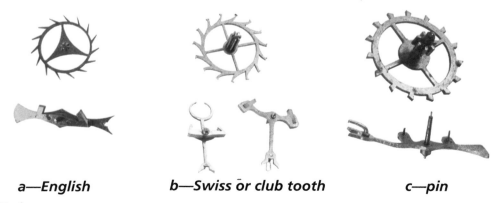

a—English   **b—Swiss or club tooth**   **c—pin**

Figure 4-27—lever escapements.

The Swiss or *club tooth* lever **b** has the ends of the escape wheel teeth flattened into a foot. Usually, the pallets are set at right angles to the line of the lever. A straight line can be drawn from the escape wheel arbor to the balance arbor running through the lever arbor. Some club-tooth levers have the right angle arrangement of the English lever and the shape of the lever varies. Both the English and Swiss levers use jewels for the pallets.

The *pin lever* escapement **c** uses two steel pins on the lever instead of pallet jewels. The escape wheel has wide, stubby teeth with the faces inclined. It can be either a straight line or right angle lever (as shown here).

Figure 4-28 shows the escape wheels of verge, cylinder, and American duplex escapements. All three are very distinctive, and all have teeth rising up from the base of the wheel; the verge teeth are parallel to the arbor (**a**), the cylinder teeth are raised up on stalks (**b**), and the short teeth of the duplex are bent up at an angle (**c**).

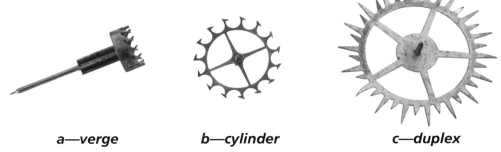

a—verge   **b—cylinder**   **c—duplex**

Figure 4-28—escape wheels.

One thing that you should always do when holding a running watch is to *look at and listen to the escapement*. Escapements behave differently and sound quite different, and it is often possible to guess the type of escapement by listening to it without even looking at the movement.

We should also determine other features, such as the number of jewels, type of balance, the form of the balance spring, the type of regulator, and the type of keyless work.

Escapements and balances, although based on abstract principles, are essentially examples of practical mechanics. There are hundreds of different escapements and hundreds of different designs for balances and balance springs. Some of these have been created to increase the accuracy of clocks and watches, but some seem to be whims rather than genuinely useful. Although all are based on the two principles I have described, they are constructed and act in very different ways.

Probably the best book for beginners that includes escapements is Cutmore's *The Pocket Watch Handbook*. I also recommend Britten's *Watch and Clockmakers' Handbook, Dictionary and Guide* and Meis's *Pocket Watches From the Pendant Watch to the Tourbillon*.

The book by Meis has a very good photographic section, but it also has a 46-page introduction that examines technical aspects of watches, including a good section on escapements. The only problem is the translator who produced the English text knew nothing about watches! Consequently, the text is strange and at times difficult to follow, using quite incorrect terminology. For example, he writes of the spindle escapement (actually the *verge* escapement) and says it is a returning escapement (actually a *recoil* escapement). He also writes of hook and rake anchor escapements (actually *lever* and *rack lever*). But despite these problems, the book explains escapements quite well, and the whole of the text section is well worth reading carefully. (*Wristwatches, History of a Century's Development* by Kahlert, Muhe, and Brunner was translated by the same person, and it suffers from the same bad terminology.) Because many escapements are described, I suggest you concentrate on just three to start with: the *verge* (spindle), *cylinder,* and *lever* escapements covered on pages 13, 16, and 22-25 of Meis's book.

Another book that you should read at some time, but perhaps not yet, is Paul Chamberlain's *It's About Time* (London: The Holland Press, 1978). This book contains an excellent survey of escapements as well as much other fascinating and useful material.

But remember these books are only recommendations. If you cannot get a copy to read, then there are plenty of others that might be more readily available.

Almost all wristwatches use the *Swiss*, or club tooth, *lever escapement.* There are some early watches that have the cylinder escapement, and many low-quality watches use the pin lever escapement. Also, there are a very few, very expensive wristwatches that have other escapements.

Jewels

Before discussing jewels we look at pivots.

All pivots have shoulders (Figure 4-29). If you hold a watch flat with the dial up, all the wheels and pinions will drop down until the pivot shoulders are resting on the inside of the top plate (or the cocks). If you turn it over with the dial down, all the wheels and pinions will drop until the pivot shoulders are resting on the inside of the bottom plate next to the dial. Some *end play* or freedom is necessary to reduce friction, which would be very large if the arbor was tight between the plates. There is also some *side play* and the pivots are slightly smaller than the holes they run in. If you hang the watch from your fingers with the pendant up, the pivots will rest on the bottoms of their holes, and the pivot shoulders may or may not touch the plates.

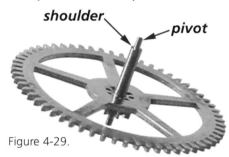

shoulder **pivot**

Figure 4-29.

The force of the mainspring, transmitted from the first wheel to the center wheel's pinion, will push that pinion away from the first wheel, holding the pivots against the sides of their holes away from the source of power. The effect of this is to continually wear one side of the hole until it becomes oval. (If you have access to an old American clock that hasn't been repaired, you will be able to see the oval holes and the large amount of side play in one direction.) Exactly the same effect happens with all the arbors of the train, but the farther the arbor is from the mainspring, the smaller is the force and so the wear is proportionally less. Wear can be minimized by lubrication and by making the pivot and hole very smooth and hard. The center arbor of a watch that is regularly serviced should show minimal signs of wear.

Friction is extremely important because there is only a small amount of power available to drive the train. For example, in one day the center wheel turns 24 times, but in one day the 4th wheel will turn about 1,440 times (once a minute), and the escape wheel 14,400 times. The balance oscillates about 216,000 times each day. By the time the mainspring's energy reaches the balance, a substantial part of it has been absorbed in overcoming friction, and the remainder has been diluted by the train so that the impulse that drives the balance is tiny. Consequently, even very small variations in friction affect how well the watch runs; the effect is most serious with the fastest moving wheels, which receive the least power.

Friction can be reduced by *capping* the pivot holes. If a pivot is slightly longer than the hole it runs in and there is a cap on the outside of the hole, then the end of the pivot will rest against the cap and the shoulder will not touch the inside of the plate.

The function of watch jewels is to reduce friction and wear. Instead of the pivots running in relatively soft metal holes, they run in very smooth, hard holes drilled through gemstones. And to further reduce friction some holes are capped with *end stones*.

Pivot holes are jeweled and capped, starting with the most delicate and fastest moving arbor, the balance staff. Then jewels are added to the arbors, working from the escape wheel toward the barrel. In addition, the escapement may use jewels; it is usual for the lever escapement to use two jewel pallets on the lever and an impulse jewel on the balance staff. In contrast, the cylinder escapement needs no extra jewels, because the cylinder forms part of the balance staff and interacts directly with the escape wheel teeth.

The slower an arbor moves, the less benefit is gained by using jewels. Provided the pivots are of a suitable size and well lubricated, there is very little point in using jewels on the barrel and center wheel arbors, and even less point capping them with end stones. Also, some arbors cannot have end stones. The center arbor extends through the bottom plate to the dial and cannot be capped; and if a watch has a small seconds hand, then one of the 4th wheel pivots cannot be capped for the same reason.

A watch is said to be *fully jeweled* if the 3rd, 4th, and escape wheel pivot holes are jeweled, the balance staff is jeweled and capped, and the escapement is jeweled. If we ignore the escapement, then ten jewels are required for eight pivots and two end stones.

This is the number for a fully jeweled cylinder escapement watch. However, a lever escapement requires an extra five jewels: two for the lever arbor pivots, two for the pallets, and one for the impulse pin. Thus, a fully jeweled lever escapement watch has 15 jewels.

The term *fully jeweled* is just a convenient way to describe the minimum jeweling expected in a good watch, and there are many variations. For example, many English fusee lever watches omit the jewels on the lever arbor. This arbor is under little stress and oscillates only a small distance, rather than rotating completely as the other arbors do. So jeweling these pivots is less important than the other pivots. Jewels were rare before the beginning of the nineteenth century, and their application grew slowly. So it is fairly uncommon to find fully jeweled watches before about 1820.

Jewels are also used as an eye-catching aid to selling watches. The Liverpool windows mentioned in chapter 3 (Figures 3-13 to 3-15, pages 40-41) do serve a useful function, but they are made huge to bedazzle the customer. Figure 4-30 is an example of excessive jeweling where the fusee arbor is jeweled with Liverpool windows, but the hidden center arbor is not jeweled! However, despite the advertising, this is a high-quality movement. The catch to hold the movement in the case is near the top of the picture. The second arm is a lever used to stop the movement running by locking the 4th or escape wheel. Although commonly called *doctor's watches*, these stop levers probably have nothing to do with timing human pulses. Most likely they were used to set the watch to a sundial. The movement would be stopped and the hands set to a convenient time. Then the movement would be released when the sundial indicated that preset time.

It is also difficult to say whether some highly jeweled American watches, such as the Hamilton 940 in chapter 3, Figure 2-43, page 22, actually *need* all their jewels or some are included to enhance the beauty and appeal of the movement. It is certainly not right to assume "the more jewels the better."

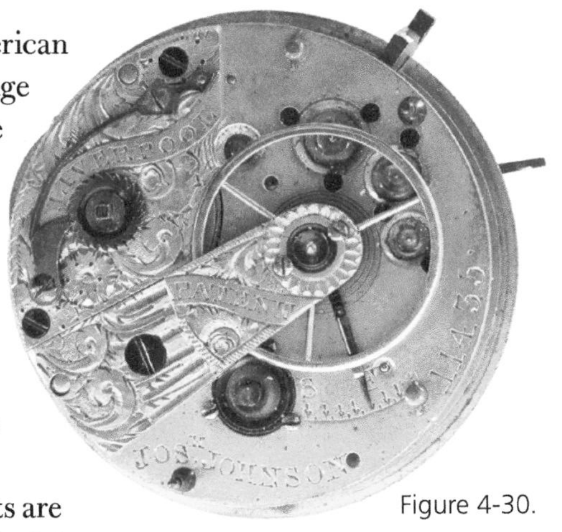

Jeweling can be deceptive. It is quite easy to count the jewels you can see on the top plate, taking care to check for end stones. So the watch in Figure 4-30 has seven visible jewels, including the end stone on the balance cock. But how many jewels are there on the bottom plate? And how many extra jewels does the escapement have?

Usually, it is safe to assume that the top and bottom plate pivots are jeweled identically, and we can simply double the number to 14 in this case.

Figure 4-30.

We can also assume the escapement has the standard jeweling for its type. A cylinder escapement has no extra jewels, a lever escapement has three extra jewels (for the two pallets and the impulse pin), and so on. Thus this lever escapement watch should have 17 jewels.

But some watches have different jeweling on the bottom plate to the top plate. This can be for practical reasons; for example, a watch with a seconds hand on the 4th wheel arbor may have two jewels for that arbor on the top plate (a hole jewel and an end stone) but only a hole jewel on the bottom plate. In other watches the bottom plate may not be jeweled at all. These cheaper watches have visible jewels to imply quality and leave out any jewels that cannot be seen and hence will not affect the opinion of the purchaser.

Accuracy

By far the most important aspect of a watch is how accurately it keeps track of time.

Two important concepts are needed to understand accuracy: First, the *rate* of a watch is the average difference between true time and the watch over 24 hours. If you set a watch accurately, then you can easily find out the difference at the end of a day. If you keep checking for several days, you can note the differences and average them. For example, your watch may be 10 seconds fast, +10, at the end of day one, +18 after two days, and +35 after three days. On each day it gained 10, 8, and 17 seconds, respectively, and so the *rate*, the average, is +11-2/3 seconds.

Second, the *variation in rate* is the difference between the rate and the actual daily variations. On the first day the variation was −1-2/3, the second day −3-2/3, and on the third day it was +5-1/3.

Provided a watch keeps time reasonably accurately, its rate is unimportant. If we set the watch accurately, go about our business for three days and then want to know the time, all we need to do is subtract three times the rate, 35 seconds, from what our watch displays and we will know the correct time. Well, nearly! Our watch does not keep the same rate every day; sometimes it runs faster and sometimes slower. Because of this variation we can only estimate the time within about five seconds.

Suppose we have two watches and the first has a rate of +6 with a variation of 0 and the second has a rate of 0 and a variation of ±6 seconds. Then the first watch is *more accurate* than the second. With the first watch we can subtract the rate multiplied by the number of days that have elapsed and, because the rate does not vary, know the time exactly. With the second watch we do not have to calculate the time because it keeps perfectly in step, but because its rate varies, we cannot know the time more accurately than within about six seconds. After four days the first watch is exactly 24 seconds fast and the second watch is up to six seconds out, but we don't know how much.

Most mechanical watches have a *regulator* to adjust the rate. This regulator alters the effective length of the balance spring and so alters the frequency at which it oscillates. By doing this, the time the balance and spring take to oscillate can be varied.

Figure 4-31 shows a *Tompion regulator*, common on pre-1800 watches; the brass pointer is rotated by a key and, at the same time, a wheel under the advance/retard dial adjusts the balance spring. Figure 4-32 shows a regulator common on nineteenth-century American and English watches; the Tompion regulator has been replaced by a simple rotating arm. In both cases the balance is said to be *undersprung* because the balance spring is between the balance and the top plate. Figure

Figure 4-31.

Figure 4-32.

Figure 4-33.

Figure 4-34.

4-33 shows an *oversprung* system. By far the most common, it is the same as the second regulator but placed over the balance. Figure 4-34 shows a *micrometer regulator* and is the same as Figure 4-33 except that the regulator arm is held firmly between a screw (marked by an arrow) and the U-shaped spring. This enables it to be adjusted very accurately.

A good survey of regulator designs will be found in Chamberlain's *It's About Time*, and Shugart's *Complete Price Guide to Watches* illustrates different regulators used on American watches.

Although the regulator can be used to adjust the rate of a watch, it disturbs the balance spring and affects the rate variation. If a watch is adjusted with the regulator in one position and then the regulator is moved, the watch may not be as accurate as it was.

Some watches are *free sprung*. These watches do not have a regulator, and their rate is adjusted by altering the balance spring and the balance. Once this is done there is no way that it can be changed without remanipulating the balance and spring. Free sprung watches are generally very accurate and are very high-quality watches.

Although it is quite easy to adjust the rate of a watch by a regulator, it is a lot harder to reduce the variations in the rate and hence improve accuracy. These variations have four basic causes: *motive power, temperature, friction,* and *gravity*.

Ideally, a balance and its balance spring are *isochronal*; that is, the oscillations take exactly the same amount of time whether the balance rotates a large or small distance. The sizes of these oscillations vary with the *motive power,* the tension in the mainspring, and a fully wound watch oscillates more vigorously than when it has nearly run down. The rotation also varies with the condition of the watch and how clean it is. In reality, different oscillations take different times, and this lack of *isochronism* causes the rate to vary.

Changes in *temperature* affect every part of a watch because metals contract, expand, and change in elasticity. Most importantly, temperature variations cause small changes in the momentum of the balance and in the elasticity of the balance spring, thus altering the oscillation time.

The *friction* between components, such as pivots and the holes they run in, varies significantly. Low temperatures cause oils to thicken and increase the resistance to motion. Over time oils break down and become mixed with dust.

Finally, the effect of *gravity* changes with the position of the watch. If a watch hangs from its pendant, then gravity will cause the balance spring to slump slightly out of shape, altering its effect, and pivots to run against the sides of their holes. But if a watch is set down with its dial up, the pivots run on their ends and shoulders, and the balance spring is pulled in a different direction. Gravity is partly offset by carefully shaping the balance spring and using an overcoil spring. But the remaining variations are still significant.

Adjusting, carefully correcting a watch to minimize rate variations, is extremely difficult and requires great skill. Very slight changes are made to the shape of pivots, the balance, and the form of the balance spring so that the watch runs equally well at different temperatures and in different positions, and the escapement is isochronal. Not only is it difficult, but adjusting a watch takes many weeks, during which time the rate of the watch is continuously checked and the variations gradually minimized.

Perfect time is never achieved. Instead, the watch is adjusted so that the rate variation is minimized for the situations in which it will be used. For example, a pocket watch is normally "dial up" when it is put down or the time is read off the dial, or "pendant up" when it is in a pocket. It is most unlikely to be placed "dial down" or "pendant down." So, after fixing the temperature and isochronal behavior, the adjuster will correct the watch first for dial up and pendant up. If it is to be a very accurate watch, it will be adjusted in other positions as well. In contrast, a wristwatch spends a lot of its time pendant down, and it is adjusted for different positions. Automatic wristwatches also generally run with the mainspring fully or partly wound. As a result, achieving isochronism with them is different from what is necessary for a pocket watch.

There is a good discussion of balances, balance springs, and accuracy in *Wristwatches, History of a Century's Development* by Kahlert, Muhe, and Brunner.

Swiss Calibers

Now we will look at some watches and apply what we have learned.

To begin with something simple, here are three Swiss bar movements (Figures 4-35 to 4-37). (The word *caliber* is used to describe a particular movement design: the sizes of the plates, wheels, and pinions, the details of the train and so on.)

Figure 4-35 is like the movement we have previously examined, which was illustrated in Figure 4-3 (page 48), except that this one is signed Baume (the forerunner of Baume & Mercier). It has bridges for the barrel and center wheel and cocks for the 3rd and 4th wheels. Then there is one other cock **e**, which runs under the balance. A single cock for the escapement fitting under the balance will almost inevitably mean a *cylinder escapement*. Swiss manufacturers produced vast numbers of these movements, and nearly all of them look the same. The movement has a plain balance and it probably has ten jewels: four for the balance and two each for the 3rd, 4th, and escape wheels.

Figure 4-36 looks like Figure 4-35, except the center wheel bridge has been flipped over. It is signed Cartier Geneve; this is probably the Cartier who founded the famous jewelry company. Again there are bridges for the barrel and center wheel and cocks for the 3rd and 4th wheels. Then there are two other cocks: **e** for the escape wheel and another marked by an arrow. To the right of the pivot jewel on this extra cock you can see a semicircle of steel. Two cocks for the escapement almost inevitably mean

Figure 4-35.

Figure 4-36.

a *lever escapement.* Note that the lever escapement is a right-angle lever; lines drawn from the balance cock jewel to the lever cock jewel and from the lever cock jewel to the escape wheel cock jewel form a right angle. The watch also has a plain balance and appears to have 15 jewels—the same ten jewels that are in the previous watch plus five for the escapement. Compare this with the Hamann watch in chapter 2, Figure 2-74, page 30, and Figure 4-41, next page; it has a straight-line lever escapement. Again, many of these movements were manufactured.

Figure 4-37 is a large watch; the first obvious feature is that it is a mirror image of the second movement. There also are two cocks at the end of the train, which might be another lever escapement, but it is not. It is a much rarer *pivoted detent escapement,* so named because it uses a detent, a lever that prevents another piece (in this case the escape wheel) from moving. It is described in Meis's *Pocket Watches From the Pendant Watch to the Tourbillon* where it is incorrectly called a rocker escapement. Although not easy to see in a small photograph, there are two features marked by the arrows at the bottom that identify it. The one on the left is a counterweight on the end of the detent, and the one on the right is a screw.

Figure 4-37. This watch has a cut compensation balance and probably has 19 jewels. The 3rd and 4th arbors have two jewels each, but the detent and escape wheel arbors also have cap jewels and the escapement requires three other jewels for its action.

The terminology for very accurate watches varies from country to country. The word *chronometer* is used rather loosely by the Swiss for any very accurate watch. But in England a chronometer (or pocket chronometer) is specifically a watch that uses the spring detent or pivoted detent escapement, and a very accurate watch that uses a lever escapement is a *half-chronometer.* In America the term *railroad watch* is used because very accurate watches were developed specifically for the railway network. Watches used for navigation are commonly called *navigation watches,* but the English also use the term *deck watches.*

Figures 4-38 to 4-40 show the escapement of these three watches in more detail; the balance cock and balance have been removed to enable the escapement to be seen more easily.

In the first two, the telltale shape of the escape wheels confirms they are cylinder and Swiss lever escapements, respectively. All that you may be able to say about the third escapement in Figure 4-40 is that it is different, but the shape of the escape wheel teeth is particular to detent escapements. The two features of the detent are now obvious: The *detent* **a** has the *locking jewel* **b** mounted on it.

Figure 4-38. Figure 4-39. Figure 4-40.

Figure 4-41 is the escapement of the Hamann watch in chapter 2, Figure 2-74, with the lever cock removed.

Figure 4-41.

Swiss bar movements have been manufactured in vast numbers over a long period of time, and dating them is not easy. A safe guess is to assume that such a watch was made circa 1880, give or take 20 years. Most are ordinary or very ordinary watches of no particular merit, but some are excellent, so do not make the mistake of assuming that all these watches are uninteresting.

Figures 4-42 to 4-44 show three different Swiss movements. The escape wheel cock of the movement on the left, hiding under the balance, indicates it has a cylinder escapement. Only the balance staff is jeweled, and it is quite possible there is only one jewel, primarily for decoration. It has a plain balance with screws around the rim to imitate a compensation balance. Figure 4-43 has a straight-line lever escapement and 15 jewels, although only four are visible in the photograph. It has a *micrometer regulator* where the regulator arm is held firmly between the screw marked by an arrow and the U-shaped spring. This allows the regulator to be positioned very precisely and suggests the movement's rate has been adjusted. From the photograph it looks like the watch has a plain balance, but it was running at the time and the balance is blurred. A quick look gives the impression that it is a cut, bimetallic compensation balance, but a more careful inspection reveals that the cuts only go halfway through the rim and the balance is not capable of compensating for temperature changes. What, at first glance, seems to be a nice movement is, in fact, a "dolled up" ordinary movement.

Figure 4-42.   Figure 4-43.   Figure 4-44.

Figure 4-44 is an example of an unexpected find. This is a three-quarter plate movement, but the plate has been hand-cut, filed, and engraved to depict a scene. There also is a small compass mounted in the plate and the case is inscribed "Chronometre." This is a very ordinary six-jewel cylinder movement, but the unusual plate and compass make it interesting and collectible. There is a similar watch illustrated in Meis's *Pocket Watches From the Pendant Watch to the Tourbillon*, so we can date this one to circa 1880.

You will find the word *chronometre* or *chronometer* used on the dials and cases of some poor-quality Swiss watches like this one. There also are many poor watches that have the word *railway* or *railroad* and pictures of trains. These are examples of the bad aspect of Swiss watchmaking. Some manufacturers deliberately created cheap watches and associated them with English chronometers and American railroad watches to fool ignorant buyers. So when you see the word chronometer on a watch, it is likely to be very, very good or very, very bad.

Roskopf Calibers

At this point you need another book in your library: *Watches 1830 to 1980* by M. Cutmore (2002). This is an excellent book that provides a history of watchmaking in England, Switzerland, and America and includes useful technical descriptions.

G. F. Rospkopf was a Swiss who designed a cheap watch by using a special caliber that he had developed. The obvious feature of these watches is they had no center wheel to make room for a larger mainspring barrel. He also mounted the escapement on a separate subplate (the *porte échappement*) so that different escapements could be used. However, most of these watches use the pin lever escapement. Cutmore provides a history of the development of this watch and illustrates the caliber design.

Figures 4-45 and 4-46 show the top plate and escapement of a Roskopf caliber. Although reasonably well finished with the steel parts polished, the plates and wheels are thin, the screws are ordinary, and the imitation jewels (the steel plates with round holes over the pivots) are extremely crude. There is only one visible jewel (on the balance cock), which is almost certainly the only one and it probably is colored glass. Note that the hand-setting square (arrow, Figure 4-45) is just above the balance cock (typical of Roskopf calibers) instead of in the center of the movement on the center wheel arbor. The escape wheel has the typical wide teeth with sloping faces of the pin lever escapement. Note the fine, concentric circles left by roughly milling the plate, which were not erased by unnecessary polishing to keep the cost of manufacturing down.

Figure 4-45.

Figures 4-47 to 4-49 show a much better quality Roskopf caliber, signed Lucida, and made by the Oris Watch Company in Switzerland. It has a display (glazed) back, and the bottom and top plate have been skeletonized so that the movement is visible. The view under the dial shows the motion work driven by a wheel on the barrel. In Figures 4-47 and 4-48 you can see the small push-piece (marked by arrows) for the rocking bar keyless mechanism (you have remembered to read Britten, I hope!). The arrow in Figure 4-49 points to one of the two steel pins on the lever **a**.

Figure 4-46.

Figure 4-47.

Figure 4-48.

Figure 4-49.

Many people have the mistaken idea that only expensive, important watches are worth collecting, and they might dislike this book because I focus on cheap, ordinary watches. But they are wrong. The impact that Roskopf had on the development of watches is vastly greater than the impact of more exalted companies, like Patek Philippe, and collecting Roskopf watches is a fascinating and popular activity. On my website is a translation of *History and Design of the Roskopf Watch* by Eugene Buffat. In 2010 over 5,000 people downloaded this book, about 25 percent of all downloads from my site and more that three times the next most popular book; clearly there are many people who collect these watches.

The last watch, Figures 4-50 and 4-51, is not a Roskopf caliber but a wristwatch with a conventional pin lever movement. The German company Kienzle seems to have made a specialty of pin lever movements, although it did produce higher-quality watches. The pin lever is normally associated with low-quality watches, but this automatic wristwatch has 17 jewels and has pretensions at least of being of reasonable quality. Some Swiss makers used the pin lever escapement to produce small 15-jewel wristwatches for women.

Figure 4-50.

Figure 4-51.

Wristwatch Calibers

Early wristwatches were derived from small pocket watches. Figures 4-52 and 4-53 show a wristwatch in a silver case with a hinged back, but no inner dome. The movement is similar to the Swiss bar calibers we have looked at. The major difference is that the barrel bridge has been replaced by a plate to accommodate the keyless mechanism winding wheels. It is easy to see that it uses a straight-line lever escapement.

The silver case is hallmarked (Figure 4-54), but these marks are different from the usual English hallmarks. In fact, they are English marks, but the case has been imported and a special "import" mark is used. The assay office is London, and the case was made in 1916, thus confirming this is an early wristwatch.

Figure 4-52.

The luminous hands and numerals of early twentieth-century pocket and wristwatches used radium paint and can be highly radioactive; I have a tin that contains luminous hands, and a radiation expert refused to open it because it was too dangerous to do so. Small, loose flakes of the luminous material could be inhaled and result in lung cancer. Treat such watches with great care!

Figure 4-53.

quality

import mark

date letter

Figure 4-54.

The watch in Figures 4-55 and 4-56 is of a similar date and also has a red "12" numeral. But its quality is quite different, with a cylinder escapement and possibly only one jewel. The anchor trademark is for Numa Jeannin.

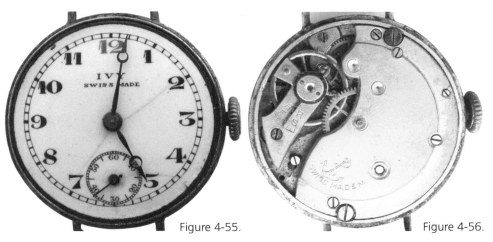

Figure 4-55.

Figure 4-56.

Figure 4-57.

The movement in Figure 4-57 is also very similar to the Swiss bar caliber pocket watch. By looking at the escape wheel cock that fits underneath the balance we can deduce that this is a cylinder escapement; the plain brass balance also confirms this.

Figures 4-58 and 4-59 are of a later wristwatch. This design, with the central bridge supporting three pivots (center, 3rd, and 4th wheels), is very common. The wire loops for the bracelet indicate this watch is quite early.

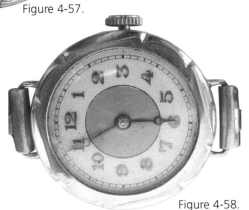

Figure 4-58.

Figure 4-59.

Although the layout varies, the movements in Figures 4-60 and 4-61 still have the same basic structure, modified to suit the shape of the calibers. Figure 4-60 is signed "Rotherhams," a major English pocket watchmaker in the second half of the nineteenth century that was reduced to distributing imported Swiss watches in the twentieth century.

Figure 4-60.

Figure 4-61.

Figures 4-62 and 4-63 show two later movements. The Cyma watch in Figure 4-62 (made by the Tavannes Watch Co.) has progressed from wire loops for the bracelet to lugs with fixed bars; it dates from the 1930s and has a divided half-plate movement. Figure 4-63 is signed "Consul," but under the dial it is stamped with the trademark of Fabrique d'Horlogerie de Fontainemelon, one of the Ebauches S.A. companies.

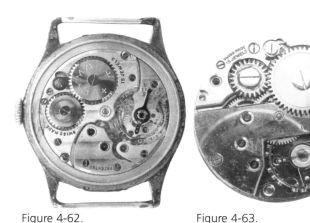
Figure 4-62.   Figure 4-63.

Figures 4-64 and 4-65 are views of an Election watch, also with fixed bars in the bracelet lugs. Although a large watch, the movement is quite small and has a divided three-quarter plate.

Figures 4-64 and 4-65.

Figure 4-66.

Figure 4-67.

Figure 4-66 is a Roamer watch. In this case the maker's trademark is visible (Figure 4-67), stamped on the inside of the bottom plate near the balance. MST is the company Meyer & Studeli, and this is their caliber 414. Unlike the other watches with makers' marks that we have seen, where the maker and signature belong to two different companies, "Roamer" is a signature of Meyer & Studeli.

The next two watches (Figures 4-68 to 4-70) are both Omega Seamaster movements. Figure 4-68 is inscribed "To Glenn Love Lorraine 14-3-70," which dates the watch; despite its high-quality movement it is a scrap watch, so the otherwise devaluing words don't matter. Both watches have the same basic movement, caliber 565, but Figure 4-70 is an automatic. This automatic mechanism is a separate unit added onto the basic watch.

Figure 4-68.   Figure 4-69.   Figure 4-70.

Both the Omega watches shown above and the Titan in Figures 4-71 and 4-72 have screw backs; the Titan has 14 flats around the edge. Both also have bracelet lugs for separate spring bars. The automatic mechanism of the Titan is an integral part of the caliber and not a separate unit.

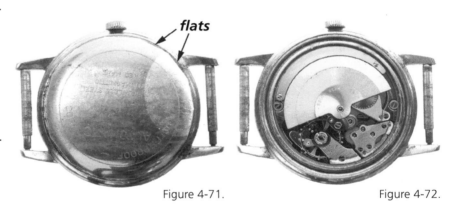

flats

Figure 4-71.   Figure 4-72.

Most of these wristwatches have straight-line lever escapements. Some wristwatches will be found with cylinder escapements, and many cheap movements use pin lever escapements. But wristwatches with other escapements are very rare.

Although there is some variation in design, wristwatch movements are quite uniform in appearance. Other than differences in quality there is little to distinguish them.

Complications A number of different complications can be added to a watch.

Alarm

Alarm mechanisms are simple and useful, but they are surprisingly rare. Most alarms have a third hand in the center of the dial, and this hand is set to the time the alarm is to sound. Most alarms also have the alarm mechanism under the dial where it cannot be seen without pulling the watch to bits.

The word *cadrature* (derived from the French word for dial, *cadran*) is used for any underdial mechanism. So the cadrature of an ordinary watch is the motion work.

The watch in Figures 4-73 and 4-74 is unusual because the alarm mechanism is mounted on the top plate, where it is visible, and it is set from the back. The back of this watch is hinged at the pendant so that it can be opened and the watch stood up on a table. Inside the back there is a glass cover that protects all the movement except the setting wheel **a**. By turning the setting

Figure 4-73.

Figure 4-74.

wheel **a** the hand on the alarm dial **b** can be set to the required time. When the alarm goes off, the hammer **c** strikes repeatedly against the wire gong **d**, which surrounds the movement. On the right of the pendant (top arrow in Figure 4-73) there is a push-piece for setting the hands. On the left there is a small lever. This watch has two barrels and two mainsprings: one for the time train and one for the alarm. The pendant normally winds the time train barrel, but by moving the lever it will wind the alarm mainspring.

The alarm mechanism (Figure 4-75) is quite simple. A wheel **c** with a disk **b** is mounted friction tight on the center wheel arbor so that it can be turned independently of the center arbor but normally rotates with it (just like ordinary motion work). The disk **b** has a notch cut in it (arrow). The long lever **a** has one end resting on the disk **b** and the other end sitting in a slot in the pin **d** on the hammer, preventing the hammer from dropping onto the gong. At the chosen time, the end of the lever at **b** drops into the notch, and the other end of the lever moves away from the pin and allows the alarm to sound. This arrangement of a lever preventing another piece from moving is called a *detent*, and a lever like this is an integral part of spring and pivoted detent escapements.

Figure 4-75.

Although difficult to see in the photograph, the disk **b** has the inscription "Breveté S.G.D.G." There is also a trademark, which I have not been able to trace. Just as some people think *Aiguilles* is a watch company, many people think a watch like this is made by *Breveté*. However, *breveté* is the French word for *patent*. The bottom plate is also inscribed "Bté SGDG 20033."

The website http://gb.espacenet.com/ advanced search for 20033 publication number reveals that this Swiss patent, dated November 15, 1900, is for a "Méchanisme de déclanchement du marteau dans les montres-réveils," a mechanism for releasing the hammer of alarm watches, and was invented by Chs. Ls. Faivre. Pritchard's book *Swiss Timepiece Makers 1775-1975* tells us that Charles Louis Faivre and his son worked in Le Locle, Switzerland, and made complicated pieces.

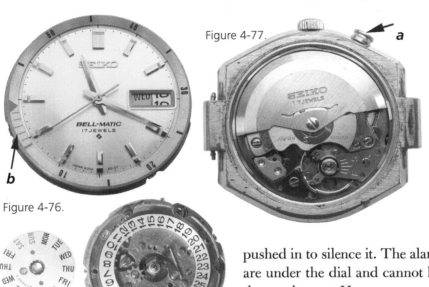

Figure 4-77.

Figure 4-76.

Figure 4-78.

Another alarm is the Seiko Bellmatic shown in Figures 4-76 to 4-78. In Figure 4-76 at **b** there is a ring labeled with hour markers, which can be rotated when the crown is pulled out; it is set to the time when the alarm is to ring, about 10 minutes to 9 in the illustration. The button **a** in Figure 4-77 is pulled out to allow the alarm to sound and pushed in to silence it. The alarm and calendar mechanisms are under the dial and cannot be examined without pulling the watch apart. However, even after removing the dial, alarm ring, and day ring, the alarm mechanism cannot be seen. The only visible part is the wheel that meshes with the alarm ring (arrow, Figure 4-78).

Chronograph

A *chronograph* is a normal watch, showing the time of day, with an additional mechanism that measures short periods of time. A *stopwatch* only measures time periods and does not show the time of day. The central sweep-seconds hand and one or two minute and hour registers can be started, stopped, and reset to zero at will. Many people call chronographs *chronometers*; please don't do that!

Chronograph mechanisms are almost always placed on the top plate, as in Figure 4-80. This chronograph is quite simple. It has a single button in the crown to start, stop, and reset the mechanism, and a 30-minute register: the small dial at XII, see Figure 4-79. But even so, the view of the movement (Figure 4-80) shows that this is a very complex mechanism. Having only one controlling button

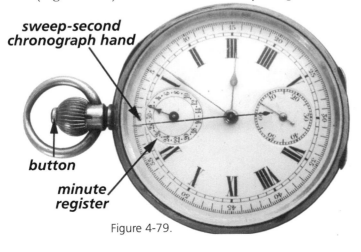

sweep-second chronograph hand

button

minute register

Figure 4-79.

Figure 4-80.

means that the actions of start, stop, and reset are repeated in that order, and it is not possible to stop the hands and restart them without first resetting them.

Despite being complex, chronographs are quite common. They certainly have had many uses—from timing sports races to timing scientific experiments and even controlling space vehicles. However, their popularity is partly because of the complex dial, which provides the wearer the opportunity to show off the sophistication of his or her watch.

The wristwatch chronograph shown in Figures 4-81 and 4-82 is different because it has two pushers at **a** and **b** (or it should, but the pusher at **b** has disappeared). The pusher at **a** starts and resets the mechanism while the pusher at **b** stops it. By repeatedly pressing **b** the hands can be started and stopped to record the accumulated time of a number of events. This chronograph also has a normal subseconds at **d** and a 30-minute register at **c**.

There are many different designs that vary significantly in quality for the chronograph mechanism.

Figure 4-81, far left, and Figure 4-82.

Repeater

A *repeater* sounds the current time whenever the watch owner wants to hear it. This is done by pressing down the bow into the pendant or by moving a slide on the side of the case. The oldest and probably most common repeater is the *quarter repeater,* which sounds the hour and the last quarter hour. The sound is made by hammers striking on a bell inside the case dome or on gongs. There are also half-quarter, five-minute, and minute repeaters.

Repeater mechanisms are almost always underdial cadratures and cannot be inspected unless the dial is removed.

Figures 4-83 to 4-86 show a quarter repeater in a two-color gold case, made in Geneva about 1785. This is a typical continental verge watch that has a jewel on the balance cock instead of a steel coqueret as shown in chapter 3, Figure 3-27, page 43. The case has a fixed back with no inner dome, and it is wound through the dial and set by a square on the canon pinion above the minute hand. The movement is hinged and held by a catch, like the English watches we have seen.

Figure 4-83.

Figure 4-84.

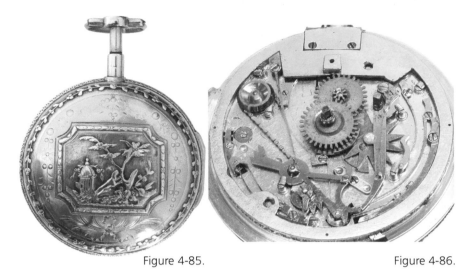

Figure 4-85.   Figure 4-86.

If you are careful, it is fairly easy to examine the cadrature of this watch. The dial is held by two feet and a small screw, which can be seen just below the number 12. Once the hands and this screw are removed, the dial will come off to reveal the repeater mechanism (Figure 4-86).

Like chronographs, a lot of repeaters were manufactured. Also like chronographs, the mechanism is very delicate and all too often has been poorly repaired.

Calendar

Calendar mechanisms are also placed under the dial and vary significantly in complexity. The simplest (Figures 4-87 and 4-88) displays the day of the month by the numerals 1 to 31. This watch, made in 1805, has a day chapter ring inside the hours and a hand points to a number. Modern watches have a date ring under the dial and one number is displayed through a window. Either way, the owner has to manually adjust the calendar at the end of each month that has less than 31 days. The mechanism is very simple. In addition to the normal motion work, there are two extra wheels **a** and **b**. Wheel **a** has a 24:1 reduction from the canon pinion so that it rotates once in 24 hours. On it there is a pin **c** that meshes with the ratchet teeth of wheel **b**. Wheel **b** has 31 teeth and is held in position by spring **d** whose head fits between two teeth; this is called a *jumper spring*. Otherwise, wheel **b** can be rotated at will because it is not connected to the train unless pin **c** is acting on it. Once every 24 hours pin **c** pushes **b** around one tooth.

Figure 4-87.

Figure 4-88.

Other calendar indications operate in the same way. A day of week indicator has the same basic design, but the ratchet-tooth wheel has only seven teeth. A month indicator has 12 teeth and can be run off the wheel **b**. Showing moon phases is a little more complicated because the moon orbits the earth in approximately 29-1/2 days and so requires more complex gearing.

A *perpetual calendar* is vastly more complex. It automatically adjusts the day of month indicator for all months, including February in leap years.

Other

There are other complications. The complications I have described are, in one way or another, useful. However, some watchmakers have created watches with features that are best described as gimmicks or displays of their prowess.

Automata are moving figures or objects. A simple automaton is a watch with a painted scene on the dial. The 4th wheel arbor, which normally has the seconds hand on it, can have attached a miniature waterwheel or windmill that rotates every minute. A more complex automaton is a repeater with painted figures standing beside bells. When the repeater is activated, the figures move and appear to strike the bells.

Chiming watches sound the hours (and quarters) automatically. Because this would drive people mad, they have a silencing facility.

Musical watches have a built-in music box and play a tune on demand.

Erotic watches have an erotic scene painted on them or have automata enacting sexual activity. Such watches are illustrated in Carrera's *Hours of Love*.

Finally, some watches include more than one complication. Watches that have both repeater and chronograph mechanisms are quite common. Much less common are *grande complications*, which include everything possible in one watch. Although a wonderful display of the watchmaker's skill, they are primarily toys for the extremely wealthy.

How Do They Work?

Both Meis's *Pocket Watches From the Pendant Watch to the Tourbillon* and *Wristwatches, History of a Century's Development* by Kahlert, Muhe & Brunner give some information about complications, but I recommend a different book: Lecoultre's *Guide to Complicated Watches*. This book provides very good, clear descriptions of different complications and their mechanisms. Carrera's *Hours of Love* gives some information about automata mechanisms. And if you are seriously interested in repeaters you should read Watkins' *The Repeater*.

However, a warning is needed. Many complications are hidden under the dial and cannot be seen, and those on the top plate, like chronographs, are very complex. These mechanisms vary considerably in design; consequently, it is very difficult to assess a complicated watch, and beginners should try to avoid them until they believe they have enough knowledge and experience to be able to assess their condition and quality. Even then, watches with complications sometimes have to be bought "blind" without being able to judge the state of the mechanism properly. In particular, although a quick test may suggest the complication works, more thorough investigation can reveal serious problems. Unfortunately, many complicated watches have been repaired by watchmakers who did not understand the mechanisms, and often serious damage has been done. So the basic rule is: Do not buy a complicated watch until you have a very good, detailed understanding of how it works and what faults to look for. Otherwise, you will inevitably waste a lot of money.

Chapter 5

The Condition Game

Figure 5-1.

Marriages, Fakes, Replicas, and Frauds

I have already discussed the problems with makers and signatures when we first looked at watches. But the problem of who made a watch is even more complicated when it has been rebuilt, modified, or carries a false signature. These watches can be roughly grouped into marriages, fakes, and replicas.

Marriages. A marriage is a watch made up from separate components belonging to different watches, most often cases and movements.

Pocket watch marriages are rarely to deceive and nearly always to protect or enhance a movement. A simple example is Figure 5-1. Although hard to see, the arrow points to a flattened part of the case edge where a screw once held a movement. At some time the original movement had been taken out and replaced by the present one. Actually, I know this happened a few years ago, because I did it. This nickel case had a high-quality American movement in it and an ordinary American movement was in a nice gold case, so I swapped them.

Such marriages of a movement with a different, not original case are quite common with American watches, which were made to standard sizes. But some care is needed before pronouncing a marriage. The very first watch we looked at was an American movement in an English case, but it is not a marriage.

Marriages are much less common with English watches because standard sizes were rarely used in England and cases were handmade to fit a particular movement. So finding a case to fit a movement can be very difficult. Sometimes the marriage is simply to protect an interesting movement.

Americans regard swapping cases to be a sin. But, strictly speaking, we can regard most American watches as marriages, because cases and movements were sold separately, and if a different customer had entered a shop, the same movement may well have ended up in a different case.

Figure 5-2.

Figure 5-3.

Figure 5-4.

The key wound and set movement in Figures 5-2 to 5-4 was made about 1860 by Sir John Bennett, London, and has an unusual "resilient" escapement. This escapement uses two semicircular springs to limit the movement of the lever instead of fixed banking pins. To protect it, it has been put in a modern, nickel stem-wind case, and the back has been replaced by a perspex (a hard, transparent plastic) cover with winding and setting holes. The movement is a good fit, but part of the case had to be cut away to make room for the balance.

Such marriages of convenience are done not to deceive but to improve or preserve and are rarely misrepresented. However, the collector needs to be aware of the possibility and carefully evaluate any dubious watch.

Figure 5-5.

The watch shown in Figures 5-5 to 5-7 is a good example. This quarter repeater was purchased off the Internet. Less information was provided by the seller than I show here, and the case was described as circa 1750 sterling silver with the implication that the movement dated from the same time, when it actually is circa 1800. The incorrect bow and push-piece immediately sounded alarm bells and the movement had definitely been recased. This does not mean it is a bad buy, just that caution is needed. Once I had the watch in my hands the marriage became clear. The case is, in fact, hallmarked Chester 1909 and is a standard nineteenth-century English case with the winding hole in the dome neatly sealed. The movement is too small for the case, and it has been very skillfully fitted to it. A bottom plate with its hinge (probably from the original movement) has been cut out to form a ring that fits tightly around the bottom plate of the repeater. At Figure 5-6a you can see the hinge of the repeater's bottom plate inside this ring, and **b** is that ring's hinge. In

Figure 5-6.

Figure 5-7.

the dial view you can see that the dial is too small. The movement catches had to be removed from both bottom plates, because there is not enough space for them, and a screw has been neatly inserted to lift the movement out of the dome. I don't know if the seller knew any of this, but it is quite likely he did.

Both of these marriages are sensible solutions to the problem of preserving a bare movement. There are many very good, uncased movements around, the result of melting down gold cases to sell the metal. The care and effort in recasing these two movements show that a previous owner respected the watchmaker's work.

The watch in Figures 5-8 to 5-11 is an interesting example. The movement is a 17-jewel Massey lever with a stop lever **b** and signed Robert Perry. I have not found Perry listed in the reference books, and the name may be a trade name used by another maker, but the movement must be circa 1840 because of its style. However, the case is a massive, smooth silver case that is clearly not English. Most obvious is that there are no hallmarks and the case

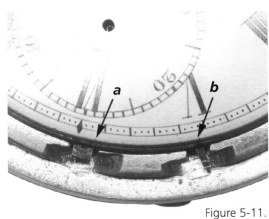

Figure 5-8.

Figure 5-9.

maker's mark "MB" is not a known English maker. Less obvious, but equally important are that the bow is sprung into the pendant, whereas English bows are held by a screw, and the movement catch **a** and the stop lever **b** fit into small cutout spaces on the wide edge of the case, whereas all English cases have thin case edges (Figures 5-10

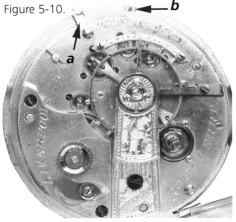

Figure 5-10.

Figure 5-11.

and 5-11). It simply isn't an English case, but there is no doubt that the case was specially made for this movement, because of the perfect fit and because it bears the serial number of the movement.

However, there are two American casemakers who might have used the initials in Figure 5-9, Margot Brothers of Boston and Mathey Brothers of New York, and both worked in the 1860s. (My source for this information is the 2002 edition of Charles Crossman's *A Complete History of Watch and Clock Making in America*.) So it is likely this is a marriage and the original case was removed (because it was damaged beyond repair or melted down) and a new case was made.

Of course, the ideal is to fit a movement in a contemporary case. The above examples are obvious marriages to anyone who can estimate the movements' dates from their style or who has checked when the makers worked. But when a movement is recased using a contemporary case it can be difficult or even impossible to detect the marriage, or even to be sure there is a marriage.

In contrast, wristwatch marriages are almost always to deceive. Some wristwatches are assembled from a mixture of genuine and other parts to produce a watch that appears to be made by a particular company but is just a collection of bits. For example, a genuine Rolex case and dial can be used with a cheap ébauche. Such a watch can be very difficult to distinguish from the real thing, and Richard Brown in his book *Replica Watch Report* gives them the very appropriate name of *Frankenwatches*. The difficulty collectors have in looking at wristwatch movements is a serious disadvantage in such situations.

Fakes. A fake is a watch that is deliberately signed with the name of a famous or important maker, but it is usually obviously inferior, at least to someone with knowledge about watches. Fakes make no attempt to look like the real thing and rely on ignorance for their success. A simple example is a $10 quartz watch in an ordinary case but with the dial signed "Rolex." It looks nothing like a genuine Rolex. There are also many fake wristwatches being made in China with brand names such as Breitling and Vacheron Constantin, although some of these may be better described as replicas.

Many fakes are good enough to fool the ignorant. I once bought a $50 Rolex Oyster, which looks real when seen on someone's arm. But hold it and examine it and the fraud becomes clear; it feels wrong and, although having the right features of a Rolex case, nothing is quite good enough or accurate enough. There is no need to open the case and see the poorly finished, cheap movement to decide it is a fake. However, it requires some experience with genuine Rolex watches to recognize these faults.

An interesting wristwatch in this context is the Jules Jurgensen shown in chapter 2, Figures 2-67 to 2-69, page 27. Although this is not a fake, having been "made" by Jurgensen, a person who bought it expecting a movement as good as the signature implies might be bitterly disappointed in the same way as the purchaser of an actual fake.

There are many fake pocket watches. For example, one of the greatest watchmakers was Breguet, who produced quite unique and exquisite work. But you will also find many ordinary watches with his signature on them—mainly plain, standard verge watches. Such watches were manufactured (probably in Switzerland), engraved with the master's signature, and sold cheaply to anyone who would buy them. However, there has been more than one watchmaker with the name Breguet, and the signatures on some watches may be genuine, just the wrong Breguet.

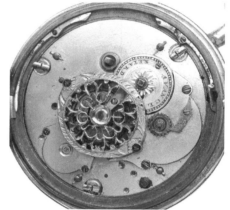

Figure 5-12.

Another example is the Dutch arcaded minutes watch signed Wilders (see chapter 3, Figures 3-11 and 3-12, page 40) which looks nothing like an English watch but has an English signature.

It can be very hard to decide if a watch is a fake, especially if you want it to be genuine! Figures 5-12 and 5-13 show the movement and cadrature of a verge quarter repeater. This is the remains of a nice watch. The silver case is original, but the dial is a poor replacement and the watch has several faults. However, it is very nicely made.

The case has an inner hinged dome (Figure 5-14) made of gilt brass and signed "Berthoud a Paris." Is the watch made by Ferdinand Berthoud, his illustrious nephew Louis, or someone else? The quality suggests it is made by Berthoud, but the signature on the dome is off-center! I am afraid I find it hard to believe either of the Berthouds would have had their name engraved off-

Figure 5-13.

Figure 5-14.

center, and I suspect this watch is a fake, although it is unlikely that I will ever know. But it doesn't matter much because the watch is a nice example of its type.

To some extent "time heals" applies to fakes. A 200-year-old fake watch, like the one shown above or the Wilders watch in chapter 3, Figures 3-11 and 3-12, page 40, is interesting and collectible in its own right, whereas a modern fake Rolex is just a nasty watch.

Replicas. A replica is a watch that deliberately sets out to deceive and appear to be the real thing, even under close inspection. Unlike fakes, which are made to fool the ignorant, replicas aim to fool people with knowledge and, very importantly, make lots of money for their creators. The Berthoud watch we have just looked at could be regarded as a replica, but pocket watch replicas are uncommon and most replicas are wristwatches. A serious replica is one I heard of recently. An apparently genuine Patek Philippe chronograph worth around $100,000 was offered for sale, but an astute person realized that a Patek Philippe watch with the same serial number had sold recently. The replica could only be recognized as such because of the duplicate serial number and differences in the fine detail of the movement. Both used the same Valjoux ébauche, and the people who made the replica had used sophisticated automatic tools to copy the case. Although it would have cost several thousand dollars to create the replica, the potential profit was considerable.

A good example of both fakes and replicas are watches signed Tobias. Earlier I suggested that the Tobias Company may have imported Swiss watches and exported them to America under its own name. But the reality is more complex. Tobias took some English makers to court and proved that they had been making replicas—similar English watches falsely signed Tobias. There also is good evidence that Swiss watchmakers produced fake Tobias watches. Some were crude imitations of English-style watches, and many were ordinary Swiss bar movements. But these watches do not rule out the possibility that Tobias also distributed such watches. To complicate matters, Tobias used other English names on his second and third grade watches, including William Robinson! So deciding if a watch signed Tobias is genuine or not can be very tricky. (For more information see Michael Edidin's "English Watches for the American Market" in the October and December 1992 *NAWCC Bulletin*s.)

Frauds. Of course, not only are watches manipulated to increase their value but some makers have used what could be politely called deceptive packaging. They rely on the ignorance of the customer and the inability of the ordinary person to open a watchcase and look at the movement.

The watch shown in Figures 5-15 to 5-17 could be called a "character watch" in more ways than one. The screw back case houses a "Zorro Musketeer deluxe 17 jewlis Zorroflex" movement, and the back announces "The Thomas Cup Sport Watch 1939." But the Swiss-made movement is a low-quality pin lever with one to four jewels on the balance staff, and its style suggests it was made in the 1950s. Well, the "17" is ambiguous, what "jewlis" means is uncertain, and maybe there never was a "Thomas" cup. But there is no doubt that the dial and case are designed to deliberately deceive the innocent buyer.

Figures 5-15 fo 5-17.

Figure 5-18.

Figure 5-19.

Figure 5-18 is a badly worn dial, but it appears to read "Comet Electric ST, 23 Jewelled." Again the screw back prevents the buyer from discovering that it is another pin lever with steel plates and the informative inscription "one jewel, unadjusted, Basis Watch," Figure 5-19. Pritchard's *Swiss Timepiece Makers* informs us that Basis manufactured watches between about 1950 and 1986. Admittedly, the dial does not say "23 jewels" and I suppose "Electric" is a trade name. Of course, "Electric" might infer an electomechanical movement, in which case the fraud is compounded!

Of course, brands with a good reputation do not play this game. Or do they? Perhaps we could include the Jules Jurgensen watch in chapter 2, Figures 2-67 to 2-69, page 27, here.

Figure 5-20.

The last example is a pocket watch (Figures 5-20 to 5-22) with quite a pretty dial and an inner cover inscribed "MI Chronometre 22 Rubis." If you open it and look at the movement, there are plenty of red "stones." There are 13 adorning the cutout and engraved top plate, but whether they are rubies or pastes (colored glass) I don't know. But I do know that these are purely decorative and serve no function. The cylinder escapement movement has another four jewels on the balance staff, this time useful, leaving five jewels unaccounted for. Presumably, four are on the fourth and escape wheel pivots, but I don't know where the last one is.

I have seen many "MI Chronometre" watches, and they have all been low-quality Swiss watches with cylinder escapements, often "dolled up" to look better than they actually are and most certainly not chronometers. Another one was shown in chapter 4, Figure 4-44, page 64. I don't know what "MI" stands for, but "Modèle Imitation" seems likely.

Figure 5-21.

Figure 5-22.

Assessing Condition

Irrespective of what type of watches you want to collect, the *condition* of the watch is a fundamental and important criterion, and assessing condition is an essential part of examining any watch.

A basic point to be remembered is that *nearly every watch you come across has been used.* Now and then you might find a watch that has been kept in a cupboard or only used for Sunday best, but they are rare. If you buy a watch that is 50 years old, it has almost certainly been used for most of those 50 years. And hopefully it has been

Figure 5-23.

to a watch repairer *ten* or *more* times, for regular cleaning and servicing if nothing else. Often, but not always, repairers leave their "calling cards"; Figure 5-23 shows eight repairers' marks on the inside of a case. Most are meaningless codes and initials, but the one near the casemaker's name includes the date 26/6/47. However, many repairers do not leave marks, so the absence of marks does not mean a watch has never been repaired.

If you buy an immaculate watch, you *cannot* tell if it has had a major repair. If it has been fixed by a good repairer, there should be no sign at all that the movement has been stripped, repaired, and cleaned. So, unless there are obvious dents or other signs, or the repairer did a poor job, the watch you hold could be indistinguishable from one that had spent its life in a drawer. Consequently, we will probably never find a watch in "mint" condition. Being well used does not necessarily mean worn out or damaged. Provided a watch has been handled carefully and regularly serviced, it can still be in excellent condition.

One problem with assessing condition is: *who can I trust?* Really, the correct answer is *only yourself.* There is no doubt that a dealer with a good reputation and who will give some sort of guarantee can be trusted, but there are also dealers with apparently good reputations who may not be dishonest but may be a little careless with the truth. Similarly, there are some watchmakers who are not above creating work for themselves by telling you a movement needs repairing when it does not.

One such person sold me a very interesting watch that was in need of some repair, but the work required was well within my capabilities. It was not until I had partially disassembled it that I discovered a major, unmentioned fault; the balance staff had been very badly repaired and would have to be replaced. Maybe the dealer didn't know? Maybe he did and forgot to mention it? It is irrelevant because the watch and the problem are now mine.

Another person, a watch repairer and dealer, sold me two watches as "fully serviced," but both were dirty and in desperate need of cleaning (not to mention a couple of obvious, minor defects).

All three watches were bought overseas on the basis of photographs and dealer statements, and needless to say, I no longer buy anything from those two people.

The purpose of my stories is not to frighten you, but to stress the need for *you* to be able to assess the condition of a watch. Many watch sellers know little about watches, and you have no choice but to make up your own mind. Those who do appear to know about watches have to be assessed carefully. Certainly, nothing beats holding a watch in your hands and examining it yourself, but in these days of international transactions and the Internet, that is often not possible. If you see a very interesting watch on eBay, do you refuse to buy it because you cannot handle it? Sometimes, but often you will take a risk based on your knowledge and your desire.

Working Watches

First we had better decide what working means. Working means the watch is in a *good, usable condition*. It doesn't just tick, but everything about it works. It might need cleaning, but it does not need to be repaired.

To decide if a wrist or pocket watch works, you really need to have it for a couple of days, but the basic checks only take a few minutes of looking, listening, touching, and smelling:

1. Wind it. Try to stop the seller winding the watch before he gives it to you; you want it completely run down if possible. Wind it half a turn or so, then look and listen. Is it sluggish? Maybe it won't run unless you shake it and then it soon stops? Hopefully it will start easily and run reasonably enthusiastically. Listening is important. A watch in good condition sounds good. Also, each escapement makes its own particular noises that help you assess its condition. Wind it up, listening to and feeling the action of the keyless mechanism as you do so. Then look and listen. How vigorous is the escapement action?

2. "Position, position, position" is a basic rule for buying real estate. It is also a very good rule when buying watches! If it runs, turn it upside-down and listen; or hold it above your head so that you can see. Then hold it sideways and look and listen. (I have a nice fusee lever watch that runs vigorously as soon as you wind it. Turn it upside-down and it promptly stops!)

3. Set the hands. Listen to the keyless work, feel if it is smooth or rough, and see how the hands move. Turn the hands through more than a complete hour in case they touch somewhere. If there is a small seconds dial, leave the hour hand over it, and let the watch run for a minute to see if the hour and seconds hands touch. Touching hands is often trivial and usually only requires the hands to be manipulated a little. But occasionally it is a serious problem; for example, if the minute hand wavers up and down as you turn it around, then the center wheel arbor may be bent.

4. Smell the watch and, if you can, look carefully at the movement. Does it smell dirty? If so and the escapement is sluggish, it might just need a good cleaning (but it might need much more). Can you see scratches, burred screw heads, mismatched screws (e.g., one is brightly polished and the rest are blue), missing screws, or other signs of poor care?

5. See how clean it is and if there is any rust or other corrosion.

6. Enamel dials often have hairline fractures and occasionally chips out of them, and metal dials are frequently discolored; chipped enamel dials and badly discolored metal dials are very difficult to repair satisfactorily.

7. And don't forget to carefully check the case! Covers should be firm and fit well. Dents usually mean the watch has been dropped. If it has been repaired since it was dropped, it may be "as good as new" or it may still have problems.

Wristwatch movements are generally in good condition because their cases seal them from dirt and moisture better than pocket watches, and more recent ones have built-in "anti-shock" jewels. But because you often have to buy them without seeing the movement, you must be extra careful during your examination of what you can see.

Although this examination will tell you most of what you need to know, it will not tell you everything. Ideally, you should wind a watch and keep it running for a few days to test how well and accurately it runs. It is very difficult to detect problems caused by wear without a detailed test and examination.

If the watch runs well, the only other thing you can check for is obvious signs of minor damage.

The three common movement faults are missing or incorrect screws, scratches, and corrosion. Figure 5-24 shows two screws (marked by arrows) that are blued and have a polished ring, but a third screw is a different shape and simply blued. One of the original screws on the barrel bridge has been broken or lost by a previous repairer and has been replaced. Although these three screws look rather ordinary in the enlargement, with the naked eye two are quite beautiful and the odd one stands out.

Figure 5-25 shows quite bad corrosion. Again, with enlargement the corrosion and dirt are much more visible. Most of the corrosion is superficial, affecting the appearance of the watch and not its running. But the rust on the balance spring regulator may mean the balance spring itself has rusted, which would be a disaster. Also, it is not possible to see if the arbors and other internal parts are rusted without partially disassembling the movement. But this is a particularly bad example for illustration, and I would hesitate before classing such a watch as working, even though it actually runs quite well.

Figures 5-26 and 5-27 show an example of utterly incompetent repair work. Around the five plate screws (marked by arrows) there are deep gouges into the plates exposing the underlying brass; clearly, a screwdriver that was much too large has been used. Damage like this cannot be repaired, and the condition and value of the watch have been permanently degraded.

Figure 5-28 highlights scratches from previous repairs. Scratches caused by slipping screwdrivers are quite common (and indicate carelessness), but I have no idea what the repairer was doing to produce the long scratch under the balance. Pinned movements like this one often have marks around the pillars. The scratches marked with an arrow on the left are caused by the repairer inserting a pin and then cutting the end off with a knife! Not only that, it seems he rested the knife blade on the regulator scale and scored it very badly. Quite often one or more pins will be missing. In

Figure 5-24.

Figure 5-25.

Figure 5-26.

Figure 5-27.

Figure 5-28.

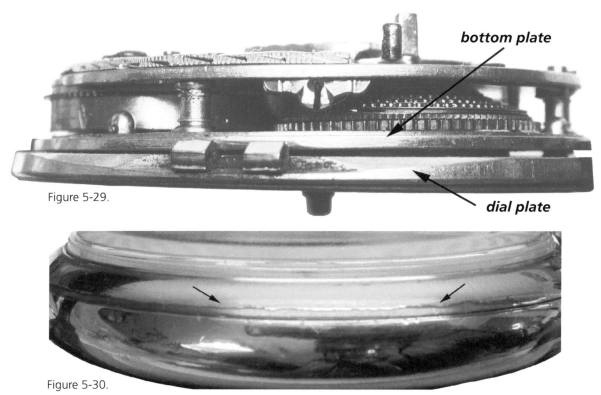

bottom plate

Figure 5-29.

dial plate

Figure 5-30.

Figure 5-29 the enamel dial is pinned to a plate, the dial plate, and then this plate is pinned to the bottom plate of the movement. The gap between the two plates shows that at least one of the pins is missing.

Figure 5-30 is a warning! When I saw this American Rockford watch, it was obviously clean, running well, and had a very nice movement. It wasn't until I got it home that I checked the case. This is a *gold-filled* or *rolled gold* case consisting of brass with a thick layer of gold on both sides. It is immaculate except for part of the edge of the bezel where the gold has flaked off. Because polished brass and gold are quite similar in color, a casual glance will not detect this fault.

Sick and Borderline Watches

Dealers and antique shops sell watches that work, but they also sell watches that "work" but are less than perfect. Also, dealers, antique shops, and watchmakers quite often have some seriously sick watches "under the counter," and these can be well worth examining. But it should be assumed that if the dealer couldn't fix them, then you must regard them with skepticism. However, because they are a liability, you may be able to buy them cheaply.

Sick watches have the advantage that for a relatively small outlay you might be able to get an interesting and repairable timepiece. But then, you may not! So it's a good idea to approach buying such watches as gambling. The basic rules of gambling are that you can afford to lose your money and you will have a lot of excitement and fun doing so! But gambling on watches is rarely a total loss, because you can use them to learn about watches and watch repair and later you can strip them and keep some of the bits for spare parts.

Sick watches are much harder to assess, but they can be roughly divided into three types:

Yuk! this is junk. The movement is worn out, in poor condition, and may have broken parts.

Oh, well, this is junk. The watch may be repairable, but it is not worth the effort.

Oh dear, this watch is sick. The watch is repairable and it would be worth the effort.

The basic question is: would the total cost of the watch and the repairs exceed the value of the watch? I am not referring to profit or investment, but something much more basic: *If you buy a sick watch and repair it, the total cost cannot be more than the price you would have to pay for a similar working watch.* So what a person who knows nothing about watch repair classes as junk may be viewed quite differently by someone who can undertake some repairs and so rejuvenate a watch for a much smaller financial outlay.

Some faults can be corrected easily and others are very difficult and expensive. Indeed, many sick watches are not worth fixing up, but this may not matter if their value lies in what you can learn from them. Figure 5-31 is one such watch. It has no case, no dial, and no fusee chain, and it is in bits. There also doesn't appear to be enough wheels! However, this movement is still rather nice despite these defects; provided it didn't cost too much!

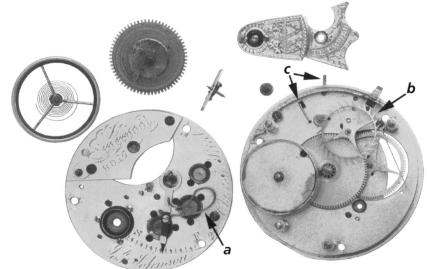

Figure 5-31.

Like the watch in Figure 5-8, page 77, it has a *Massey lever* escapement; and no wheels are missing because it has a three-wheel train. The escape wheel **b** is much larger than normal and has a "seconds" hand mounted on it. The regulator to adjust the balance spring and the rate has a *bimetallic curb* **a**. After the watch has been adjusted, this curb reacts to temperature changes and changes the rate by altering the curb pins. The balance is blued steel with a rim of gold applied to it, and the movement has 15 Liverpool windows; if it had a normal four-wheel train it would have 17 jewels. Finally, it is a "doctor's watch" where **c** is the stop lever; moving the lever causes the fine wire arm to press against the escapement lever and stop the watch. As I mentioned in chapter 2, a doctor's watch probably had nothing to do with medicine, but the stop lever was to enable it to be set accurately to a sundial.

Actually, except for the problem of finding a dial and a case, this movement is not all that sick. All the important parts (in particular, the pivots and balance spring) are OK, and it needs little more than a good cleaning to run well.

It must be remembered that watches are machines and, like cars, they will eventually wear out. How long this takes depends on how well and frequently a watch is serviced and how many "crashes" it suffers. Any watch, no matter how old or how good, can reach the point where it must be sent to the "wrecker's yard" and scrapped. But as with cars, sometimes the remains found in a field or a drawer may be worth expending a lot of time and effort to restore. So before a watch is scrapped, you should carefully assess whether it really is junk or whether someone might still want to restore it.

A lot depends on the quality of the movement. Some cheap watches were manufactured from poor materials and constructed in a way that makes repairing them almost impossible or simply too expensive. They are a bit like some quartz watches where it may be cheaper to throw away a watch and buy another one than it is to replace the battery.

You should also remember that as a collector you have a museum, and objects in a museum may have considerable value even if they are incomplete or faulty. So even a cheap, unrepairable watch may be an important example of one aspect of watchmaking.

Figure 5-32.

Figure 5-33.

The watch shown in Figures 5-32 and 5-33 is less hopeful, but it does have a dial. Indeed, the unusual dial with the owner's name, Burt Robinson, for the hour markers is the main feature, although the movement is good, being a 17-jewel English lever. However, the movement is missing the end stone from the balance cock, the fusee chain, *and* the escapement lever! Furthermore, if you look carefully at the dial, you can see that the hole for the seconds hand has been elongated. Sad to say, this proves the movement and dial are already a marriage!

Well, one marriage deserves another! I might see if I can fix the movement but, because it is not particularly interesting, I may be better off finding another movement for the dial; perhaps my "in bits" Massey lever. In the mean time this sick watch can languish in a drawer.

The rusted movement in Figure 5-25, page 83, is a more realistic example of a sick watch. This is a nice 23-jewel watch in a silver hunter case, and despite being filthy, it runs quite well. Now, if there is no hidden damage, then it should be possible to clean the movement and repolish the steel caps on the cocks. It is likely that some marks will remain and the watch will never be pristine again, but it would still be a nice example of this type of watch. However, if the balance spring or other important parts have rusted, it may well not be worth trying to repair it.

Besides dirt, which often is of little importance because the watch will be fine after a good cleaning, there are two very common faults:

The watch winds forever and will not run.

This fault occurs with going barrel watches and simply means the mainspring has broken so that no power is transmitted to the train. When the spring breaks, a sudden jolt is transmitted to the train and can cause some damage to the rest of the watch. Often if you apply a *little* pressure to the barrel you can get the watch to run a bit and so find out if the problem is limited to the spring. The important word is *little*. If the balance doesn't swing with a little pressure, it is tempting to apply more force, but the most likely consequence of this will be even more damage. However, such a test is not perfect and some damage may not be detected; for example, the sudden jolt can bend or break the teeth of a wheel. Mainsprings break quite often and are easily replaced, so provided the watch is in good condition otherwise, it is not serious.

To avoid the damage caused by mainsprings breaking, many American watches have a "patent pinion." The pinion on the center wheel arbor, which meshes with the barrel teeth, is screwed onto the arbor. If the mainspring breaks, the jolt will cause the pinion to unscrew from the arbor and so prevent damage. Of course, this assumes that the watch has been serviced properly and the pinion hasn't stuck tight after years of neglect.

If the watch is pendant wound, also check the hand setting. Occasionally, a watch will be fine and the mainspring good, but the keyless mechanism has broken and the watch cannot be wound or set. Sometimes a faulty keyless mechanism is easy to fix, but it can be difficult, so regard such watches as very dubious. An example of this problem is given below.

In the case of fusee watches, a broken mainspring will result in a loose or broken fusee chain, which can jam the fusee and make it impossible to wind the watch. If the fusee is not jammed, it may "wind forever" like a going barrel. Fusee chains sometimes break when there is nothing wrong with the mainspring, and they exhibit the same symptoms. Because fusee watch movements are nearly always hinged to the case and can be swung out for inspection, it is easy to check for a broken chain. Again you should see if you can apply a little pressure to the train to find out if the movement will run.

The watch is overwound.

This common expression is completely misleading, because it is not possible for a watch to be overwound. If a watch is fully wound and does not run, and it is not simply too dirty, then there is almost always some catastrophic damage.

By far the most common cause is that one or both pivots of the balance staff have broken off and the balance is unable to oscillate. For most watches a new balance staff has to be made and this repair is both difficult and expensive. The same effect can occur if a pivot of one of the other wheels has broken, but other pivots are stronger and less likely to break.

Indeed, a broken balance staff is one of the most common problems, because the pivots on it are very small and delicate and must cope with the weight of the balance. A sudden shock can give the balance so much sideways momentum that the pivots bend or break, which is why wristwatches usually have shock-resistant jewels for the balance staff. These jewels are held in place by springs that can move when there is a sudden jolt, thus allowing the pivots to move instead of breaking; see chapter 2, Figure 2-69. page 27.

Failure to run when fully wound can be the result of other problems. For example, it is possible for the lever of a lever escapement to move to a position where it locks the balance. This is easy to detect because the balance swings easily but will not oscillate; you can only move it one way and when it moves back, it stops suddenly. Often this indicates a fairly serious repair is needed, but occasionally it is easy to fix by removing and replacing the balance with the lever correctly aligned. However, always assume the worst.

To be sure of the cause of such faults requires more than a visual inspection and, for the first time, the tweezers you have been carrying become important. But remember your first lesson and be *very careful* or you might cause more damage. Using a loupe and tweezers, you can very gently push and lift the balance, and broken pivots are then obvious. If you prod a good balance, it will not move sideways and you can only lift it a tiny distance; but prod hard enough and you will promptly break a pivot! With a broken pivot the balance can be waggled about. When there is a broken mainspring or fusee chain, you can use tweezers to apply a *little* pressure to a train wheel and see if the train and escapement will run.

Although I have said it is not possible to overwind a watch, I am not being strictly correct. What I mean is that it is not possible to overwind a watch and for it to stay in that state. But it is possible to fully wind a watch and then try to wind it some more. If you do this, the extra pressure that you apply to the crown or key will be transmitted to the train and can cause instant damage. As soon as you release the extra pressure, the tension transmitted to the train will revert to the normal amount provided by the mainspring, and it is only while you are holding the crown or the key that the term "overwinding" has any meaning.

Having done this myself I can give you a good example of what might happen. I once had a watch that worked quite well, but the winding action was very stiff. Rather than fixing the problem I foolishly just wound the watch. Because I was not sure if it was fully wound, I gave the crown an extra twist. Lo and behold! The hands started turning at high speed, then they slowed down, stopped, and the watch ran normally. I tried again and the same thing happened. What I had done was weaken the joint between the center wheel and its arbor enough for the arbor to turn (with the motion work and hands) while the wheel was motionless, held by the rest of the train and the escapement. I hadn't completely broken the joint, but severely weakened it and, of course, it had to be repaired.

Examples of other illnesses.

Figure 5-34.

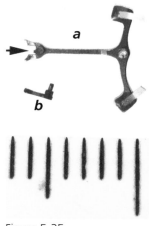

Figure 5-35.

Figure 5-34 is a sick watch that is well worth repairing. This is a complete Hamilton master navigation watch. Basically, it is in excellent condition, but it keeps stopping. The problem is that the escapement lever **a** (Figure 5-35) should have a small pin, called the guard pin, sticking out at the arrow, but it has broken off. Fortunately, some spare parts are still available despite it being 60 years after the watch was made, and **b** is a replacement guard pin waiting to be put on the lever, after which the watch will be fine. (The lines at the bottom of the photograph are part of a ruler divided into millimeters. The lever is a bit over 6 mm long.)

Although this Hamilton watch can be repaired "like new," it may definitely not be like new. This movement is engraved "Adj. Temp. and 6 Positions," confirming that *when it was made* it was very carefully adjusted to have a good rate with little variation. But what is its rate now, after 60 years? Hopefully, if it has been handled with care, its rate should still be very good, but it is impossible to guarantee this. For example, if the balance staff has been replaced, then the repairer may not have been able to, or simply did not bother, to readjust the watch. Also, the accuracy of a watch will vary with normal wear. I would still expect the watch to be very good, but it may not be as accurate as when it was first made. So any claims about the accuracy of a secondhand watch must be dubious unless it has been properly tested and readjusted.

Unfortunately, spare parts are often unattainable. Figure 5-36 is the keyless work for a medium quality Waltham watch. When I got the watch, it was in very good condition in a very good case. The only thing was that I couldn't wind it or set the hands! Ever hopeful and ever willing to take a gamble, I bought it assuming I would have no trouble repairing it. This is a simple shifting sleeve mechanism. With the crown in, the *setting spring* **a** holds the castle wheel up against the winding wheel. Pull the crown out and the lever **b** pushes the setting spring **a** down and the castle wheel engages with the motion work. The lever **a** and the part **c** of the spring are shaped so that they lock together and the crown has to be pushed in to revert to the normal winding position.

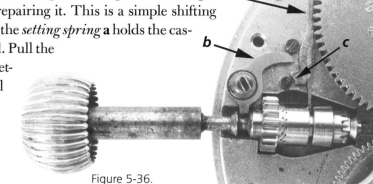

Figure 5-36.

This is fine in principle, but the end of **a** going into the slot in the castle wheel had snapped off and the castle wheel was just flopping about. I suppose a new spring could be made, but it would be a lot of work and too expensive; the watch simply isn't worth that much. Rather than give up and trash it, I went through my collection of bits and pieces. There I found several setting springs like **a** but also broken; until now I hadn't known what such a spring should look like and didn't know I was keeping worthless junk! But in amongst them I found a complete spring of the right size, and a few minutes later I had converted a junk watch into something I could use or sell.

Actually, a lot of spare parts for watches are still available, many years after the companies making them have gone out of business. Watch repairers, material suppliers, and other people have kept "hoards." The problem is to find the person who has the piece that you need.

Such spare parts were only produced after watchmaking reached the precision necessary for parts to be interchangeable, which is generally after about 1930. Certainly, it is impossible to get replacement parts for any watch made before about 1860, and there are few parts for any watch made between then and 1930. So repairing old watches inevitably involves making or adapting parts to fit.

The last watch, Figures 5-37 to 5-39, is a bit more difficult. Except for the badly discolored plastic crystal (easily replaced) and one deep scratch, the case is very nice and is engraved with the logo of the American Brotherhood of Locomotive Firemen and Engineers. But the movement, in addition to some corrosion on the winding wheels, has a broken balance staff and the original micrometer regulator has been replaced with a simple regulator; the gold "star" on the balance cock is all that is left of the original regulator. Although potentially a nice watch, we again have the problem that the repairs may exceed its value. Being an American watch, I managed to get spare parts to fix it, but it also languishes in a drawer until I get around to repairing it. Provided I am careful it should be possible to return this movement to near new condition, although it may no longer be properly adjusted.

These examples of sick watches have been chosen because they all had potential when I got them, but I have plenty more that do not have any potential and are simply a source of spare parts; most have broken pivots and I lack the skills necessary to make new balance staffs. Into which group a particular watch falls depends on your *interests, capabilities,* and *wealth* and, of course, whether the total cost is no more than a similar working watch would cost. If you are interested in a watch because it illustrates some feature of horology, you might not bother getting it fixed. Otherwise, do you have the skill to repair such a watch or do you have the money to pay someone else to overhaul it? Or, like me, will you be happy to leave such a watch in a drawer to await some future time when it might be fixed? After all, I should not scrap a watch just because I cannot repair it, unless I am certain it doesn't deserve a better future.

Figure 5-37.

Figure 5-38.

Figure 5-39.

All these examples have been of pocket watches, mainly because I only collect pocket watches. But also, if you come across a sick wristwatch, what do you do if you cannot open the case and look at the movement? Again, the wristwatch collector is restricted by the design of these watches and has no choice but to take a blind gamble or simply refuse to buy anything that is not in good working order.

These examples just scratch the surface. A watch can have one or more of hundreds of different defects, and the condition game, like the others, must be played every time you pick up a watch.

The Repair Game: Strip Jack Naked!

Now, be honest! Did you read the last section about sick watches? Or have you decided only to collect good, working watches from reputable dealers, so you skipped that bit to get to this interesting sounding section faster?

Sorry, but you can't do that! No matter what you collect you are going to need some understanding of watch problems. You may never pull a watch apart, let alone repair one, but without at least an understanding of what faults watches can have, you are likely to end up in trouble, and collectors who know nothing about watch repair are restricted in what they can sensibly collect. Here are two reasons.

The first reason is that without dismantling a watch and examining it carefully, it is impossible to detect all the faults it might have. Let me give two examples.

The Hamilton 940 watch that I have shown in chapter 2, Figures 2-43 and 2-44, page 22, as far as anyone can tell, is in excellent condition. Its dial and case are perfect, there are no marks on the movement or incorrect screws, and it runs faultlessly. Any collector who wanted one of these watches could not buy better.

Unfortunately, if you did buy it, you would be inheriting a nasty problem—a problem that is invisible until the movement is pulled to pieces, and even then it might not be spotted. The jewel in the dial plate for the 4th wheel pivot is cracked. And if the watch was wound and run regularly, this crack would almost certainly cut into the pivot and eventually ruin it. I know this fault exists only because I have completely stripped and cleaned the movement and happened to spot it.

The other example is an old Rolex Oyster watch I bought at an auction. I didn't know why, but no one seemed to want to bid for it, so I got it very cheaply. The case and the crystal looked in need of attention and the screw down crown was not right, but it seemed to run OK. So maybe I had gotten a bargain.

Nope! When I took off the back, I found the automatic rotor and the inside of the case had a thin film of rust! What I hadn't realized was the back had only been screwed on hand-tight, and the people who did not bid had taken the back off and seen this mess! Important lesson: *Always check everything you can when you examine a watch.*

However, all the other people who had seen this mess had only looked skin deep. A more careful inspection revealed the rust to be only on the case and rotor (the movement itself was clean), and it was little more than a superficial film. After getting Rolex to refurbish the case and cleaning the movement myself, I had a cheap, working watch that performed as well as a new one (I carefully tested it on a timing machine). Unless the back was removed to reveal a slightly discolored rotor, no one else would know what had happened.

This sounds like a success story, but it was only half a success. When I stripped and cleaned the movement, I found a broken part, a spring with three arms that had lost one arm. Just to see what would happen, I reassembled the movement with the broken spring and found that it worked without any apparent problems. So I left it like that. No one will know unless I am honest and tell them or they strip the watch down.

The second reason is that the collector who knows nothing about repair is restricted in what can be bought unless watches are to be kept "as found" or a watch repairer is paid for improving them. Some of my sick watch examples could not have been repaired if I had to pay someone to do it, because the total cost of the watch and the repairs would have exceeded its value. To repeat what I have already said: If you buy a sick watch and repair it, the total cost cannot be more than the price you would have to pay for a similar working watch. So when you choose a watch to add to your collection, you better be very sure it doesn't have hidden faults. And the only way you can do that is to have a good understanding of what can go wrong and what symptoms to look for. Even then it is likely that you will end up with the occasional unpleasant surprise.

Let me remind you again—nearly every watch you come across has been used. If you buy an immaculate watch, you cannot tell if it has had a major repair (e.g., a new balance staff). Unless there are obvious dents or other signs or unless the repairer did a poor job, the watch you hold could be indistinguishable from one that had spent its life in a drawer.

So if you haven't read the last section, go back and read about sick watches before continuing.

Repairing watches is not something a beginner should attempt lightly. But at least some knowledge of how watches are repaired (and made) is not just desirable but necessary, and a little experience will not go astray. But before I discuss watch repair, consider Figure 5-40. At the bottom is part of a metric ruler and the picture is about 3/4 of an inch wide. The wheel on the left is about 9 mm in diameter and a tooth on it is about 0.3 mm wide, roughly 0.012 inch. The fusee chain on the right has links that are the same as those of a bicycle chain, but they are less than 1 mm or 0.04 inch long. The hook at the end is about 1.5 mm long and about 0.1 mm thick. These pieces are from a pocket watch and are quite big. If you examine a wristwatch movement, you will find the wheels and other parts are much smaller.

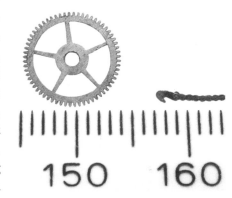

Figure 5-40.

Now, imagine I give you a file, a disk of brass and a tiny bit of steel and ask you to file a tooth on the disk and make a fusee hook on the bit of steel. Perhaps you have excellent mechanical ability and can do so, but I suspect many people are like me and would find the task next to impossible. Certainly to succeed would require much practice, skill, and a very small file. Just contemplating the problem gives me a feeling of great admiration for people who used to make watches by hand.

In fact, parts like these have not been made by *hand* for about 200 years. They have been made using special tools. Wheel-cutting machines quite precisely cut teeth in wheels. Very small punches were used to punch out the individual links that make up a fusee chain (but they were then riveted together by hand). Steel wire was drawn through dies to form pinions and other shaped pieces. And so on.

There are three essential requirements to repair a watch: *patience, planning,* and more *patience.* It is simply not possible to be a successful watch repairer if, like me, you are impatient. Or, like me, you can develop the basic skills and do many tasks yourself, but you will continually come to grief and be beset by unnecessary problems. Sometimes I need to pick up a small screw, but I am too impatient to select the right tool and use what is at hand. As a result, I drop it and spend the next hour hunting around the floor looking for it. Or I unscrew a piece carelessly and a small spring flies up, hits my glasses, and vanishes behind my bench. One of my worst experiences was when I was cleaning an automatic wristwatch. I removed the necessary screws and lifted off the rotor only to have 20 or so tiny steel balls run all over my bench! It was then that I realized the rotor ran on a ball race with the balls sandwiched between the rotor and a rim. As several balls headed for the floor and freedom, I hoped the watch would "make do" with a few missing! So, unfortunately, I know that no matter what I do, my personality will prevent me ever calling myself a competent repairer.

And, finally, everything you do must be carefully planned and done in a correct sequence—like a watch I cleaned recently. I pulled it apart, cleaned it, checked all the parts, reassembled it, and put it back in its now polished case. With a sigh of satisfaction I turned the crown to wind it. Nothing! I had done everything slowly and carefully except I forgot to check the mainspring. Oh well, go through the whole process all over again and this time fix the problem and hope that I don't forget something else!

Understanding How

Now, at this point there are two possibilities: You only want a bit of knowledge about watch repair and don't intend doing it yourself, or you are really interested in the possibility of fixing sick watches.

If you only want to understand "how to" without actually doing it, then I recommend two good books:

Fletcher, *Watch Repairing as a Hobby*. This was first published in 1947 and has been reprinted several times. Fletcher provides an introduction to disassembly, cleaning, and assembly and then describes some basic faults and their repair. The book is very well written and the instructions are excellent.

Whiten, *Repairing Old Clocks and Watches*. Like Fletcher, Whiten has written a book for the amateur who only has a few screwdrivers and a kitchen table, and again the basics are covered excellently. However, Whiten goes further and discusses more complex repairs. I am not entirely happy with these parts of the book, but if you only want an education rather than practical experience, then my reservations don't matter.

At this point I should note that there are some books I think you definitely should not buy and read because they are mediocre, or even bad. Rather than go into details I refer you to my bibliography *Mechanical Watches*; although it includes my personal opinions, it is a good guide to what is and is not worth reading.

Even if you don't intend repairing any watches, you should still pull one or two to bits and, hopefully, get them back together again.

To do this, all you need are tweezers and some screwdrivers, as shown in Figure 5-41, **a** and **b**. A set of watchmaker's screwdrivers consists of ten different sizes with interchangeable blades. There

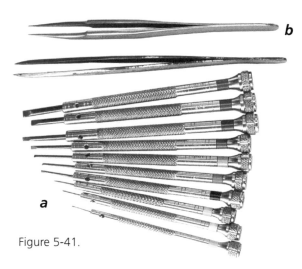

are so many simply because you *must* use the right size blade for each screw; remember the damage shown in Figure 5-26, page 83. However, the small sets of imitation watchmaker's screwdrivers that can be purchased from hardware and craft shops will do to start with. In addition to the tweezers **b**, a hand remover **c** is very useful, but not essential (how it works is obvious when you hold one).

Figure 5-41.

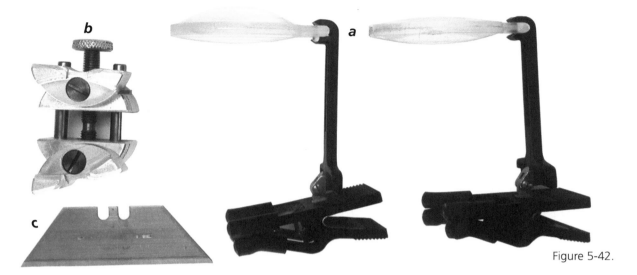

Figure 5-42.

As shown in Figure 5-42, a loupe, or preferably two of different magnifications (Figure **a**; mine are designed to clip onto my glasses), a movement holder **b** is handy so that a watch plate can be firmly gripped while you work on it. You can simply hold parts in your fingers, but fats and acids from them will leave marks; occasionally, you will come across a watch with a perfect fingerprint of the previous repairer on it. Instead of a movement holder you can use finger cots made from squares of clean, white paper to protect the parts. I suspect making *finger cots* might be a forgotten art, so here is how: Fold a small square of tissue paper to form a triangle. Fold over the base about 1/8 inch and continue until you have made three or four tight folds (or you can roll it up and flatten it). Put your index finger so that the tip occupies about half of the remaining space. Fold the top of the triangle over the top of your finger and bring up the two sides around the finger and twist together. Make a couple more for your thumb and middle finger.

In addition, if you intend tackling an old watch with pinned plates, then a sharp, fine knife blade **c** is useful to get the pins out; this is called a "Stanley" knife blade in Australia. But be gentle! A quick twist of your wrist and a tiny brass pin will disappear over your shoulder and never be seen again. Or, if you press too hard, you will cut the end off the pin.

Finally, you need a suitable place to work. A proper watchmaker's bench would be nice, but almost anywhere will do. It needs to be clean and have a *very good light source.* Preferably, it should be quite high, or you should use a very low seat; you need to have your nose close to what you are doing. A piece of smooth, white photocopy paper is about the size of the area you need to provide a clean surface to work on. With just these tools, and after you have read Fletcher or Whiten, you can strip and rebuild almost any simple timepiece. It is important that simple improvements (e.g., replacing hands and screws) can be done by anyone, and all the tools in Figures 5-41 and 4-42 will be useful even if you never try to repair a watch.

To begin with you should tackle one or more junk watches so that if you damage or lose a part, it will not matter too much. But your aim must always be to pull apart a watch and reassemble it *without any damage* so that it looks and works exactly the same or better after you touched it. Of course, this can be a bit frustrating and it would be nice to be able to clean the movement as well.

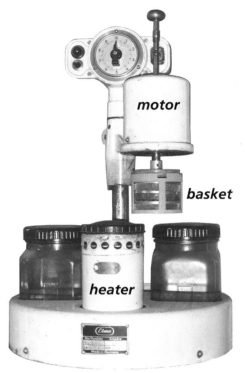

Figure 5-43.

Cleaning is messy, unpleasant, and smelly unless you have some sort of cleaning machine. Although it can be done by hand, an ultrasonic cleaner with three or four jars or an old mechanical cleaning machine (Figure 5-43) is very desirable. This mechanical cleaner has three jars holding cleaning and rinsing solutions and a heater for drying. The parts are put into a basket, which has three small, partitioned subbaskets made of wire mesh, and the basket is attached to the motor spindle (Figure 5-44). The frame carrying the motor can be rotated around the center pillar so that it is over a jar or the heater, and the basket can be raised and lowered as required.

Cleaning, like any other watchwork, requires care to ensure parts are not damaged. In particular, if you use water-based solutions, you must be very careful when drying the parts to avoid rust. This, together with some watch oil and watch grease, is all you need to work on reasonably good, working watches. But even these watches may have a few minor faults, particularly missing, broken, or poor screws and missing or unsuitable hands.

In addition to starting with junk watches so that you gain some experience without risking damaging a good watch, these watches can be stripped down and some of their parts used in other watches. However, you will quickly discover that screws come in a wide variety of sizes with different threads, and other parts may look similar but are often different. Hopefully, you are sensible and do not try to force any screw into a hole, doing permanent damage in the process; take the time to find the right one. In the process you will learn that replacing screws in later watches is much easier than in early watches; later watches are to some extent standardized, whereas early watches are often handmade and almost unique. Likewise, you will find hands come in an amazing variety of sizes and shapes, and finding a suitable replacement hand can be very tedious. Figure 5-45 is a small collection of screws, hands, case parts, and fusee pieces gathered over a few years, which is an invaluable resource for repairing other watches.

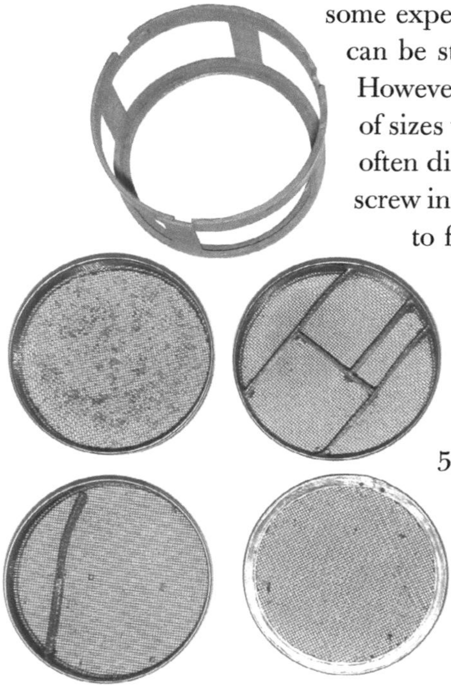

A small amount of practical experience will quickly give you an appreciation of the dexterity and skill required and will teach you a lot about what is necessary to make and repair watches. The biggest sin is to do something that a later repairer cannot correct or has very serious trouble correcting. So if you do undertake some repairs, then there are two fundamental rules: (1) If you cannot repair a fault properly so

Figure 5-44.

Figure 5-45.

that the watch is basically the same as it was when first made, then *whatever you do must be reversible.* (2) If you modify a watch movement, then it must be possible to remove the modification and convert it back to its original form.

For example, the movement catch for the repeater in Figures 5-5 to 5-7, page 76, has been removed and lost, but it should have been kept with the watch so that a later repairer could put it into a more appropriate and better fitting case. Another example is a watch I have where the dial is missing two of the three feet used to attach it to the dial plate. Eventually, the feet should be remade, but in the meantime I have glued the dial to the dial plate. I made sure that the glue can be easily dissolved and removed, and I will keep a note about it with the watch.

Ordinary glues and two-part epoxy glues, which are especially nasty, should be avoided wherever possible. Some authors suggest using them and UV glues, but such advice should be ignored. Where necessary, watchmakers use a "glue" called shellac. This softens when heated, enabling the parts to be manipulated into their correct positions, and it can be removed very easily because it dissolves in alcohol. It is far better than any modern glue.

Doing It: Books and Tools

A professional *watch repairer* is a person who has developed a wide range of skills and who does it for a living. Usually, they have spent several years studying at a horological school or as an apprentice working under a competent repairer. Like watch dealers, repairers have to earn enough to pay the rent, electricity, and insurance for their workplace and have enough left over to put food on the table. Like dealing, it is a *full-time business* that requires considerable knowledge and skill. The majority of collectors who learn how to repair watches do so as a *hobby*. Although some reach a very high level of proficiency, most remain competent amateurs and have no desire to earn a living from fixing watches, let alone accepting responsibility for their workmanship.

By far the best way to learn about watch repair is by taking a formal course of study. But it is possible for self-taught people to develop considerable skill. The two books I have mentioned are useful to read, but their limitations will quickly become apparent. Now, although there are many watch repair books, I believe there is only one really good one:

H. Jendritzki, *The Swiss Watch Repairer's Manual.* This book is brilliant and is by far the best repair book I have ever read. Unfortunately, it is not easy to find a copy, but if you can, get it and treasure it. Jendritzki describes in detail and in the correct order how to strip, inspect, and repair watches. He does this by providing excellent diagrams and photographs, and the text explains the features and processes illustrated. The word "Swiss" in the title can be ignored because the book applies to any reasonably modern watch, including American pocket watches.

There are three other books that are well worth having: (1) Henry Fried, *The Watch Repairer's Manual* and (2) *Bench Practices for Watch and Clockmakers.* These two excellent books are as well written as Jendritzki's book, but they only cover some aspects of watch repair. However, Fried includes English fusee watches, which Jendritzki ignores. (3) Bulova School of Watchmaking: *Training Units.* This book was produced as a manual for use by students in the classroom. Although it too is excellent and covers nearly every aspect of modern watch repair, the order of presentation is not really appropriate for a person learning at home.

Watchmaking and repair have been aptly described as the craft of a thousand tools. Over the last 200 years and more, watchmakers have designed and made tools for every single task they undertake.

Figure 5-46.

Indeed, if you get interested in repairing watches, you will almost inevitably start collecting watchmaker's tools as well as watches. I certainly have!

One of the most useful tools is the staking tool (Figure 5-46). This is a set of high-precision punches and anvils that are held exactly in position by the frame. It can be used for many different purposes, which are very well explained in the book *The Watchmakers' Staking Tool* by Lucchina and Perkins.

Figure 5-47 shows a few of the beautiful, small tools designed for specific watchmaking jobs, the largest of which is about three inches long. **a** is a *bow cutter*. It is used to reduce the ends of case bows so that they fit into the holes in the pendant. **b** is a *pivoting tool*. A watch arbor with a broken pivot is put in it, and a tiny drill is inserted into the frame to drill a hole in the end of the arbor in which a new pivot is inserted. It is a two-inch-long high-precision drill press! Figure **c** is a *cylinder height tool* used to measure distances between watch plates. The screw at the top is turned until the two bottom points rest against the insides of the plates, and this distance is measured from the separation of the two top knife edges. **d** is a *poising tool* that has two knife edges and is used to test and adjust the poise of balances. **e** is a *depthing tool* that is used to test and measure the separation required for a train wheel and pinion to mesh together correctly.

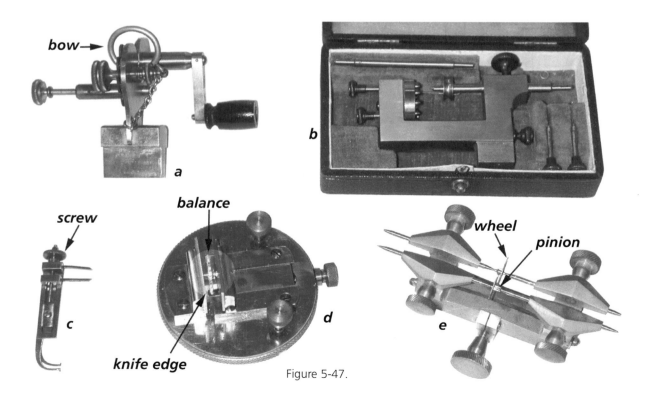

Figure 5-47.

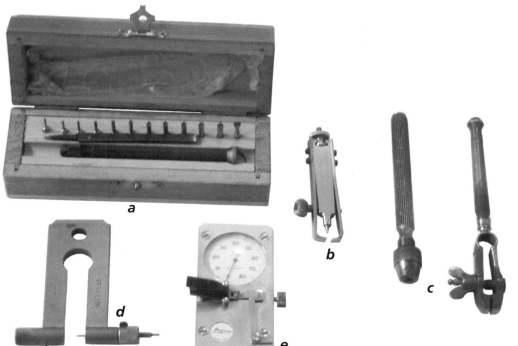

Figure 5-48.

In Figure 5-48, **a** is a countersinking tool. There are 12 cutters with pivots. The appropriate cutter is chosen and its pivot put in the screw hole. Then the cutter can be used to make a concentric countersink. **b** is a roller extractor that is used to remove lever escapement rollers from balance staffs. **c** shows two pin vices used to hold small brass and steel rods. **d** is a screw extractor. Most threaded holes for screws go through plates and cocks. If a broken screw has to be extracted, the plate is put between the two arms and two small pins are pressed against each end of the broken screw. Then the plate can be rotated to unscrew the broken piece from the plate. **e** is a special micrometer used to adjust lever escapement pallets.

Figure 5-49 shows three special pliers. **a** is a mainspring punch, which is used to punch square holes in the ends of mainsprings. **b** is a pair of bow pliers. After milling the ends in a bow cutter, these pliers are used to bend the bow to the right shape. Then it can be used to expand the bow a little to slip it over the pendant. **c** is a pair of brass-lined pliers. These are very useful for handling steel parts without any danger of marking them.

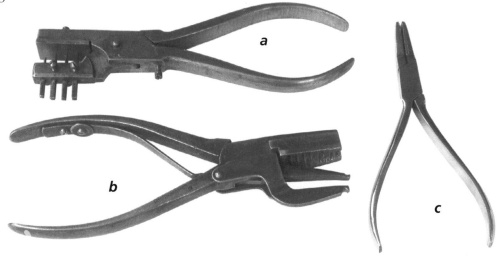

Figure 5-49.

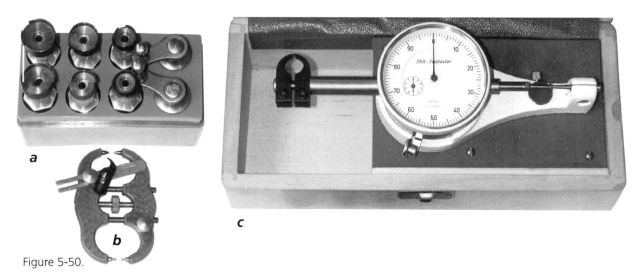

Figure 5-50.

In Figure 5-50, **a** is a mainspring winder, which is used to coil a mainspring so that it can be inserted into the barrel. **b** is a balance caliper, which is used to check the shape and truth of balances. The third tool, **c**, is a micrometer specifically designed to measure watch parts.

Figure 5-51.

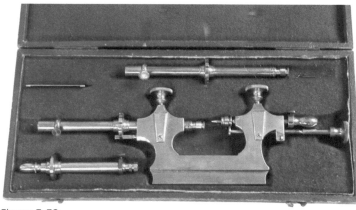

Figure 5-52.

Figure 5-51 is a *rounding up tool,* about 8 inches high, that is used to shape the teeth of wheels. In the drawer under it there is a row of different cutters for different sizes of wheels.

The *Jacot tool* or *pivoting lathe* in Figure 5-52 is about ten inches long. It is a highly specialized lathe whose only purpose is to adjust and polish pivots.

This is just a small selection of watchmaker's tools. It raises the question: What tools does the beginner need? The essentials, in addition to loupes, screwdrivers, tweezers, hand removers, and movement holders are as follows:

1. Assorted holding tools: small pliers, pin vices, a small bench vice, and brass-lined pliers.

2. Mainspring winder; see Figure 5-50a, above.

3. Crystal and case press. I illustrated one type in chapter 2, Figure 2-57, page 25.

4. Screw extractor to remove broken screws; see Figure 5-48d, page 97.

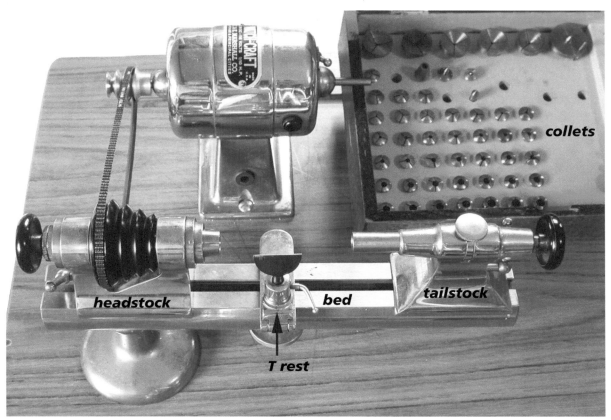

collets

headstock   bed   tailstock

T rest

Figure 5-53.

5. An ordinary *micrometer*.

6. If possible, a good *staking tool*; see Figure 5-46, page 96.

Almost all of these tools must of necessity be secondhand because some are old, and many modern tools are poor quality and not worth having.

Tools, like watches, are machines and can be worn out or in poor condition. Collecting watchmaking tools is the same as collecting watches and requires you to develop knowledge and experience. Just as you can buy bad watches you can buy bad tools. You should avoid buying expensive tools until you have a good understanding of their features, uses, and pitfalls. And you should not buy incomplete tools. Major tools (e.g., lathes and staking tools) are hand adjusted, and parts from one may not fit another or may not be aligned accurately enough to be useable.

There is a seventh tool that a beginner would be tempted to buy: a watchmaker's lathe like the one in Figure 5-53. This is a WW-style, 8 mm lathe and is probably the most common type. WW style is the shape of the bed and 8 mm is the diameter of the collets or split chucks that fit into the headstock.

However, buying a lathe must be done with care. First, the bed, headstock, and tailstock are very carefully machined to match, so that the head and tail stocks align exactly. Consequently, a tailstock bought separately will most likely be useless because it will not fit the bed properly and it will not line up correctly. Indeed, WW is a *type* of lathe, and there is considerable variation from one WW lathe to another. Some attachments are not as critical, and some mixing and matching may be possible, but it is a risky thing to do.

Second, just as the main parts vary, the collets, used to hold pieces to be turned, also vary. I have a mixed lot of 8 mm collets, some of which are too big to go in the lathe illustrated, and some are too small! But they are all 8 mm collets for 8 mm lathes.

Buying a lathe is a bit of a gamble because you have to decide what attachments you will need and are going to use. The lathe in Figure 5-53 is the bare minimum: matching headstock, tailstock, and bed, a T-rest, and a set of collets. With just these parts the lathe can be used to do most tasks. Most importantly, the tailstock holds 8 mm collets like the headstock. Lathes often come with a tailstock that can only hold small, tapered guides, but you should not consider buying one unless it has a collet-holding tailstock.

There are many books that describe tools and their use. The repair books I have recommended do so, and some other tools are explained in books like Britten's *Watch and Clockmakers' Handbook, Dictionary and Guide*. There are several books on the watchmakers' lathe and its use, but the best is *The Watchmakers Lathe and How to Use It* by Perkins.

Chapter 6

The Collecting Game

Ambitions—Good, Better, Best

If I had wanted to write a normal book, this chapter would have come first. But I have ordered the chapters on the assumption that you have already started collecting watches and advice on *how* to collect is more important than discussing *what* to collect.

Before looking at what you might collect, I want to explain what the word *best* means.

In 1993 George Daniels gave the James Arthur Lecture, the keynote lecture at the annual NAWCC seminar, and spoke on "Watchmaking in the Twenty-First Century: The Renaissance of the Mechanic." (This was published in William Andrewes: *The Quest for Longitude*, Harvard University, 1996. The book is an excellent study of marine chronometers and well worth reading.) In his talk Daniels said: "The best watches are mechanical and the most resourceful people are watchmakers. ... Of course, no one will need a watch in the twenty-first century ... but everybody likes a watch." Maybe this was a bit tongue-in-cheek, but I can't let the statement go without comment. After all, my beat-up Timex with riveted, pressed steel plates and a pin-lever escapement is a mechanical watch, but I have a sneaking suspicion that it is not what George Daniels meant! Which reminded me of a previous life ...

Before I saw the light and took up horology, I used to teach computer science—well, not science as such, there is precious little of that in practical computing, but programming. During my lectures I explained why some computer programs are better than others. Now, most people know what a car is and they can answer questions about them. So I would ask: What is the best motor vehicle? "Ferrari!" someone would yell, and then I added, "but I need one to transport 8 people." "Rolls Royce?" A bit more hesitantly. "I can't afford to buy one," I reply. "A Japanese people mover?" ...

Try again. "Ferrari!" "I like driving on beaches." "A four-wheel drive." Or "I need to move a ton of bricks." ...

The point is that there is no *best* vehicle. There may be a most appropriate vehicle for a particular purpose or a particular personality, but there definitely isn't one absolute best. If you list important criteria (e.g., petrol consumption, speed, safety, road surface, people capacity, and cost), then you will most likely get a different answer in every case. Not only that, people often impose some sort of restriction. My brother collects Morris Minor cars and for him the best car must be a Morris Minor. And it is the same with anything—the best programming language, the best dog, the best watch ...

So, if you are going to collect watches, which ones are the best? Here are a few choices, all of which are the best.

Uniqueness: I would have to plump for a Patek Philippe *grande complication*. It's a pity I don't have $4 million to spare and I will never own one, although I can lust after it in my fantasies. In reality, most people are happy with *nearly unique*. For example, a watch made by Waltham is unique in the sense that no other watch has the same serial number. It may also be nearly unique because a limited number of that model exist—the fewer the better.

Price: No problem here: a $5 quartz watch made in China, but because this book is about mechanical watches it should be a pin lever watch for the same amount. However, maybe you mean the most expensive watch? Again, price is often qualified by the amount of available money, and the best watch is one that I can afford, rather than a more nearly absolute best. This is why many of the watches photographed for this book are less than perfect.

Accuracy: Ah! not so simple, this one. I would have to choose something like my Longines VHP quartz watch, which has gained only 15 seconds in 8 years—far better than any mechanical watch ever made. But it depends on what you mean by accuracy: mean time, rate, short or long term ... So maybe a radio-controlled watch? Or an Omega with a George Daniels coaxial escapement? But that is said to be only accurate to about three seconds per month. Or a tourbillon? Or a Hamilton master navigation watch? Or ...? But who needs accuracy anyway? Most people these days manage perfectly well being within about five minutes of real time (whatever that is); if someone asks you the time, do you answer "ten to six" or "eight minutes and seventeen seconds to six"? Indeed, an old verge watch would probably be good enough for most of us. Certainly accuracy is needed, but what is meant changes with different circumstances. For example, a near zero rate would have been important for an American railroad watch, but it is not necessary for a marine chronometer.

Social status: You might want to choose a watch because it makes you look good in the eyes of others. I don't understand this. I have wandered around the streets and at parties and looked at other people's watches, but I cannot tell what they are wearing! Unless you thrust it in my face and exclaim "Breguet!" or I am very rude and try to pull it off your arm, I have no idea. But some people like to openly display their sophistication. This is why a friend bought a $50 Rolex Oyster replica from me. He could wear it, it would be impossible for an onlooker to see that it was a fake Rolex Oyster, and it wouldn't matter if he lost it. I suspect *status* is the wrong word; maybe we should call this the feel-good factor; what matters is how the owner *feels* rather than the impression made on others—in which case wearing a fake Rolex may be unsatisfactory. I like wearing my Longines chronograph even though no one notices, let alone cares.

Availability: The day I wrote this there were 453 Rolex watches for sale on eBay (although I have no idea how many were genuine). That is about 16,300 in a year! The point is that anyone can have a Rolex watch if they want one. In contrast, there were only 38 Patek Philippe and 18 Breguet watches for sale.

Lifestyle: I have a friend who is a builder. One day he wore his expensive, brand new quartz watch to work and put it in his pocket to avoid damaging it. When he got home it was not in his pocket, but he knew where it was—somewhere inside the eight-inch-thick concrete slab that he had poured! I wonder if the owners thought their house was haunted when at night they would lie in bed and hear a faint tick, tick, tick! The builder realized that he needed two best watches—one best for work and one best for going out (and maybe another for other situations).

Complications: I have never understood most complications. I don't think I have ever wanted to time something so accurately that I needed a chronograph or cared so much about knowing the time that I needed a repeater. But I do like my Longines perpetual calendar because I sometimes need to know the date and I am too lazy to correct it every month on my ordinary quartz watch. I would prefer a mechanical perpetual calendar, but I can't afford one. I know there are people who need complications, like some people actually need ten-ton trucks, but I think they are a minority.

Personality: I can ask which watch is best as an abstract concept, trying to determine some absolute, but most of the time my choice is dictated by my personality and what I want, being expressed rather vaguely by "I dunno ... I just like it." Appearance, aesthetics, and feelings matter. Indeed, many wristwatches are first and foremost items of jewelry and the fact that they show the time is almost irrelevant. In days of old, some people went to the extreme of wearing *fausse montres*, things that looked like pocket watches but were simply cases!

Brand: A couple of words on a dial can make a lot of difference. For example, a Patek Philippe chronograph could cost $100,000, but another watch using the same Valjoux movement might be only $1,000. OK, the Patek Philippe will have a much nicer case and bracelet, and the movement will have been very carefully hand finished, but are those extras worth $99,000? Of course, it has to be the right name. It might be nice to have a George Graham watch, but it had better be one made by "honest" George rather than his later namesake. Likewise, John Harrison of Liverpool made some quite nice watches, but he is not in the same class as John Harrison of longitude fame.

Social impact: Perhaps the best watch is the one that has had the most impact on society, and there is not much doubt that the honor should go to George Roskopf, although Americans will reasonably argue that Waltham deserves the title.

Age: The oldest watch or even one that is very, very old is probably unattainable—most being in museums. But maybe a watch made in 1740 is better than one made in 1940.

One final point to be kept in mind: Mechanical watches are, like vinyl records, old, superseded technology. I can buy a $10 quartz watch that will run for two or more years without attention and have an accuracy far exceeding the best mechanical watches. Brand new mechanical watches are being manufactured, but they are not produced for the general public. They are produced for collectors, investors, people who like status symbols, and nostalgia seekers.

I could go on—giving more reasons for making a choice—but you have the general idea now. And you should realize that there is no such thing as an absolute best. For some reason, probably a quirk of fate or personality, you will select some area to focus on. Your choice is arbitrary and in no way better or worse than the choices of other people—just different. Your interests also may change over time, as mine have. What matters is that the best watch for you will be *the best in the context of your chosen area of interest*.

Which is why I have recommended my bibliography *Mechanical Watches* as a basic reference. I may not have discussed the type of watches that interest you, because I have no idea what they are. But I can, through that bibliography, make you aware of books that will be relevant and useful.

Collecting Choices

Some of the alternative criteria for best that I have outlined are irrelevant to collecting. For example, watches that form part of a collection are usually not worn or chosen for their social status value. But there remain very many valid criteria on which to base a collection.

An important point is that collecting *need not be expensive*. Probably because so many books are price guides and others talk about investment, there seems to be a general feeling that the more expensive collections are *better* than the less valuable ones. This is simply nonsense. How much a collection costs depends largely on *what* you want to collect and *why* you want to collect in that area. What is important is not the *dollar* value of your collection but its *educational* value.

The thing that most interested Paul Chamberlain was the escapement. Now, it can be very hard to see and appreciate the escapement in a complete watch, so Chamberlain collected *movements*. The case, even the dial and hands, were more of a nuisance than a help. I, too, am interested in some escapements and some of my *best* watches are bare movements that don't even have to work; indeed, a few have broken pivots. I have put some of these movements into cases simply to protect them, and they are generally obvious marriages with no intention of deceiving.

Similarly, if you decide to collect examples of mechanisms, such as repeaters, isn't it better to have movements that you can pull apart and examine, rather than complete and valuable watches where you can see very little and are too afraid to pull them apart in case you damage them?

I accept that most collectors, including myself, would like to have complete, working watches, but the same argument applies. If you decide to collect Waltham watches or American railroad watches, then the cases are only important to protect the movements; ordinary nickel cases would be just as good as solid gold, but very much cheaper. The reverse is true: if you want to collect samples of American-made cases, they may as well contain ordinary, cheap movements. Some people argue that only completely *original* American watches should be collected and even switching cases is an evil. But these people don't realize that major repairs and case switches can be completely invisible and so undetectable. In addition, such watches were, in the very first place, a *marriage* of a movement and an entirely separate case chosen by the customer. If the customer had chosen a different case, there would be a different marriage.

Collectors' Books

Collecting choices depend, to some extent, on how much you know about the history of watchmaking in general and your specialty in particular.

For example, if you collect Rolex Oyster watches, then it is important to know something of the history of Rolex and the development of this watch. So I would expect that Dowling and Hess's, *The Best of Times—Rolex Wristwatches: An Unauthorized History,* and Brown's *Replica Watch Report* would be the most important books in your personal library. And if you have an inquiring mind, you will probably want to find out about its predecessors, like the Dennison waterproof pocket watchcase.

Similarly, the collector of American railroad watches needs to read Hoke's *The Time Museum Historical Catalogue of American Pocket Watches* and Harrold's *American Watchmaking: A Technical History of the American Watch Industry 1850-1930.* American watchmaking is so intimately tied to the American system of manufacture that it is difficult to separate the history of such watches from the history of mechanization. Indeed, after reading Hoke, I was so compelled to learn more that I read Jacques David's *American and Swiss Watchmaking in 1876.* And from there it is a short step to reading Daniels's *Watchmaking* to understand the modern craft.

An area that interests me is the development of the English lever watch between 1790 and 1840. Unfortunately, English watchmaking has not been well documented in books, and little has been published on this very important development that centered on Liverpool (with the rack lever in 1792, the Massey lever in 1814, and the English lever some years later). Some information will be found in the Antiquarian Horological Society catalog *Your Time.* This lack of books occurs in other very interesting areas of horology. Another neglected area is *dollar watches* (both American and European), probably because the emphasis on monetary value has caused many collectors to ignore these cheap but fascinating timepieces. Similarly, English and Japanese manufacturers (e.g., Rotherhams and Seiko) seem to have been spurned even though both produced some excellent and interesting watches, but Cutmore, in *Watches 1830-1980* and his other books, provides some valuable information.

There is a lack of information about collecting watch keys and collecting verge watch cocks (just the cocks without the watches).

However, there are some excellent books for other areas of collecting. Escapements are surveyed in

Chamberlain's *It's About Time*. And early watches are well covered by the two books by Baillie that I have already mentioned. For those who want to understand how these early watches were made Weiss's *Watch-Making in England 1760-1820* is a superb examination of the technical problems faced by early watchmakers. And Berthoud & Auch's *How to Make a Verge Watch* and Vigniaux's *Practical Watchmaking* explain how it was actually done.

Such areas of interest cannot be viewed in isolation because each is dependent on and influenced by preceding and coeval developments. Certainly, to collect watches out of context is much less satisfying than when you are able to view them in a historical perspective.

The outstanding history of the social context of horology is Landes' *Revolution in Time*. This is an excellent, readable, and easily obtained book that every collector must have. There are many general, descriptive histories but I know of no other that comes close to this work.

There are very few books that cover the histories of watchmakers satisfactorily. Most books on individual companies are very beautiful "coffee table" volumes that rarely contain useful information. Like company advertising booklets, they are based on the premise that company X is vastly superior to every other company, and this is justified by glamorizing the achievements of X and totally ignoring every development of the other makers.

American Watches

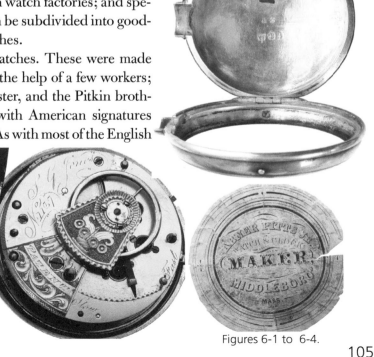

Assessing the relevance of a watch for a collection requires examining it in the context of your chosen area of interest. Many different factors, including the escapement, jeweling, finish, and maker need to be examined; such an assessment requires practice and the ability to see beyond the obvious.

The following sections look at a few areas of watchmaking and give some examples to guide you.

American watches can be divided into three types: prefactory watches made before 1850; those made in the American watch factories; and special, post-1850 watches. The factory output can be subdivided into good-quality, jeweled watches and cheap dollar watches.

There are very few prefactory American watches. These were made in small workshops by individual people with the help of a few workers; they include watches by Luther Goddard, Custer, and the Pitkin brothers. However, the majority of early watches with American signatures were made in and around Liverpool, England. As with most of the English signatures on English watches, the names are those of retailers who ordered watches engraved with their names from Liverpool makers.

Figures 6-1 to 6-4 show a typical example. Jones is a common name, but we can be confident that S. G. Jones is Samuel G. Jones, who was in Baltimore between 1815 and 1829. The movement is housed in an English silver pair case hallmarked 1817

Figures 6-1 to 6-4.

and has in it the watch paper of Abner Pitts Jr. The movement is interesting for two reasons. First, the seconds hand rotates once every 15 seconds instead of the normal 60 seconds. Second, it has a Massey lever escapement. In 1814 Edward Massey patented a form of lever escapement. It was used in some English watches until the 1830s, when it was superceded by the standard English lever escapement. It is described in Cutmore's *Collecting and Repairing Watches* and mentioned in Meis's *Pocket Watches From the Pendant Watch to the Tourbillon*. In addition to the English case, the movement is typically English, and there can be little doubt that it was made in England for Samuel Jones.

Figure 6-5.   Figure 6-6.

The watch shown in Figures 6-5 and 6-6 has been recased and much useful information has been lost. However, the dial is hallmarked 18-carat gold and has applied gold numerals, which suggests an English origin. There is no reference to James Ladd in any of the books I have checked, although there was a Ladd in Amhurst, Nova Scotia, and a watch signed H. H. Ladd, Manchester, New Hampshire, does exist. The movement is a transitional three-quarter-plate caliber like the John Wood watch in chapter 3, Figure 3-1, page 34, and was made circa 1825. Again, we can be confident that, although signed "New York," the watch was made in England.

Although the jeweled pocket watches produced in American factories may look quite different from each other, nearly all are simple timekeepers with full-plate or three-quarter-plate movements, going barrels, and lever escapements. Because there are large numbers of these watches, the beginner will have no problem starting a collection of them. We have already seen several of these watches in chapter 2, Figures 2-7, page 13, Figure 2-43 and 2-44, page 22, chapter 3, Figures 3-21 to 3-26, pages 42 and 43, and two more are shown in Figures 6-7 to 6-11.

As I have noted, early watches are full plate and later movements are three-quarter plate. The two full-plate watches shown here date from 1894 and 1886, respectively.

The watch shown in Figures 6-7 and 6-8 is a railroad watch, a Crescent Street grade made by Waltham. It has 21 jewels, but more interesting is that it is similar to the transitional three-quarter-plate English movements we have looked at. It has a three-quarter-plate layout, but

Figure 6-7.   Figure 6-8.

the top plate is a solid full plate with a sink for the balance. However, the English transitional three-quarter-plate movements have a hole cut completely through the top plate for the balance, whereas the Crescent Street movement has a sink going only about half way through the plate.

In contrast, the Illinois watch in Figures 6-9 to 6-11 has a normal full-plate layout. It has three interesting features. First, it appears to have 15 jewels: 8 jewels for the train and 7 for the escapement. However, this is deliberate deception because there are no jewels on the pillar plate and the total number is only 11.

Second, the serial number is 663,000 and this movement is the last one in a production run of 2,000 movements.

Third, the case, made by American casemaker Fahys, is unusual. When it is closed, it looks like a simple case with a hinged back and no separate bezel, which is a problem because the hands are set by the square above them and there is no obvious way to open the bezel. Opening the back reveals a hinged dome with a winding hole but, more importantly, the pendant, movement, and its protective dome are mounted on an inner, hinged silver ring. This enables the movement to swing out by the pendant for hand setting, while the balance remains protected by the closed inner dome.

Figures 6-9, 6-10, and 6-11, above.

Because their basic design is the same, collecting American factory watches depends on understanding fine details that are beyond the scope of this book. The two Elgin watches in Figures 6-12 and 6-13 were made in 1887 and 1916, respectively; the full-plate movement was earlier than the three-quarter plate, as we would expect.

They both have 15 jewels, but the B. W. Raymond (Figure 6-12) is clearly superior, because it has been adjusted and has a micrometer regulator. But how much superior? Is it a high-quality railroad watch and very collectible, or is it a fairly common example of an above average watch? Also, even though the watch in Figure 6-13 is not as good, it may be much more collectible if it is an example of a rare design. The fact that the B. W. Raymond is in a rather ordinary silver case and the other movement is in a very nice 9-carat gold case is completely irrelevant, although the novice is likely to be swayed by this factor. What

Figure 6-12.

Figure 6-13.

really matters is the importance of the movements in the context of the entire history of the Elgin company and, indeed, the entire history of American watchmaking. The point is, you as a beginner can collect American factory watches, but you should concentrate on ordinary watches until you have developed enough knowledge to recognize the important features that distinguish the important watches from the rest.

The same problem occurs with the special watches made after 1850 by watch companies and individual makers like Don Mozart. Such watches are rare and very expensive, so the beginner will probably learn about them from books long before an opportunity to buy one arises.

The Social Impact of Dollar Watches

The production of cheap watches for the masses was pioneered by Roskopf about 1865 with his *porte echappement* caliber discussed in chapter 4, but his watches are only one type of dollar watch, so named because the aim was to produce a watch that cost no more than a day's pay of about $1. We have already seen the Waterbury duplex watch (chapter 2, Figures 2-30 to 2-34, pages 20-21), but some American companies produced pin lever watches with normal four-wheel trains, and Germany also mass-produced cheap pin lever watches in the twentieth century. It is possible that Roskopf's patents prevented the use of his design in America and the American companies had to invent their own dollar watches. Certainly, they never followed the basic approach taken by the Swiss, which was to mass-produce cheap cylinder and Roskopf movements. This is understandable because the balance staff for a cylinder watch is difficult to make and quite fragile; there were large numbers of outworkers in the French and Swiss Jura producing them.

In addition to bar movements and Roskopf calibers, the Swiss produced large numbers of three-quarter-plate watches. Figure 6-14 is an example of such a watch with a cylinder escapement, only identified by a trademark of a sailor, which I have not been able to trace (Figure 6-15). The photograph of the trademark shows that the plates were left rough to avoid the extra cost of finishing them. But they are gilt to prevent corrosion.

Figure 6-14.   Figure 6-15.

Dollar watches were created by watchmakers moving in a different direction from the mainstream development of watches. They are characterized by the type of escapement and have few or no jewels. The majority use the pin lever escapement; the duplex escapement was only used in America. The cylinder watch above, which has four jewels on the balance staff, is not really a dollar watch, but it represents attempts by mainstream makers to produce competitive movements. Watches like the Waterbury would have cost more than $1, but they are early moves in this new direction. Likewise, the Lucida watch in chapter 4, Figures 4-47 and 4-48, page 65, illustrates an "upper class" dollar watch. The quality between them varies significantly, as can be seen by comparing the Lucida watch to the Roskopf caliber in chapter 4, Figure 4-45, page 65.

Collecting dollar watches is complicated by the fact that they were not made to last and many no longer work or run erratically. Indeed, the later twentieth-century movements were often constructed to make repairs nearly impossible, and some of them are the forerunners of the throwaway quartz watch.

In addition to Cutmore's *Watches 1830-1980*, this area of collecting is covered in Cutmore's *Pin Lever Watches* and Schaeder's *The Proletarian Watch*.

Early Wristwatches

Early wristwatches exhibited a variety of designs when watchmakers tried to reduce movements to the size needed for them to be worn on the wrist.

Figures 6-16 to 6-18 show an open-face pocket watch that has been converted into a wristwatch by removing the bow and adding two wire lugs for a strap. The case is silver with a hinged back and hinged inner dome, like all similar pocket watches. It was made in Switzerland and is hallmarked with the London import marks for 1912.

The watch was originally made for a woman and has an ornate dial and decorated back. It is a reasonable quality, ten-jewel cylinder movement made by "DF&C," Dimier Freres & Cie., Geneva. Very large numbers of similar watches with highly engraved silver cases and ornate dials were produced between the end of the nineteenth century and the early twentieth century.

Figure 6-16.   Figure 6-17   Figure 6-18.

Figures 6-19 to 6-21 show another converted pocket watch, but this time it is a hunter watch. Two strap lugs have been soldered to the case, and the hunter cover has been cut out so that the dial is not obscured. Again it is a ladies' movement in an ornate, gold-filled case. The movement was made in America by the Elgin Watch Company and is dated 1904 by the serial number. The hunter cover has been mistreated by not pressing in the crown before closing it, wearing away the snap so that it no longer shuts.

Both of these converted pocket watches were probably worn by men about the time of World War I.

Figure 6-19.

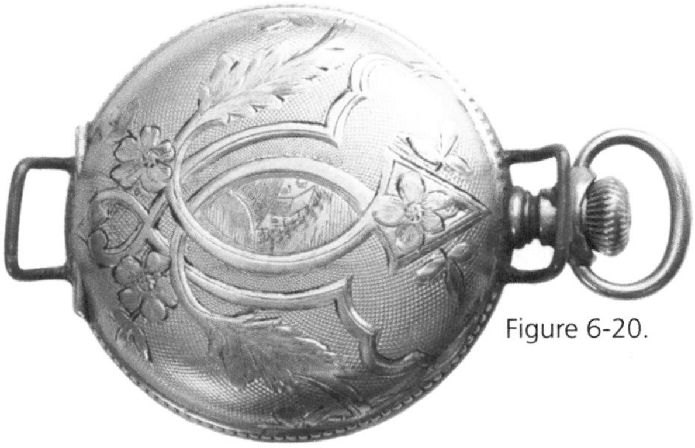

Figure 6-21.

Figure 6-20.

Figures 6-22 and 6-23 show a small wristwatch—only 19 mm (0.75 inch) in diameter—made by Baume, the predecessor of Baume & Mercier. Both the back and the bezel of the plain 15-carat gold case are hinged, and the movement is a high-quality 15-jewel straight-line lever with an overcoil balance spring. The red figure "12" is typical of wristwatches circa

Figure 6-22.   Figure 6-23.

1920. Both this watch and the one in Figure 6-16 are pin set; to set the hands the pin on the case has to be pushed in and the crown rotated. This cannot be done without removing the watch from the wrist. So, although the Baume watch was probably manufactured as a wristwatch and is not a pocket watch conversion, the movement was actually designed to be used in a pocket or hung from a pendant chain.

It is just as easy to convert a pocket watch for wearing on the wrist now as it was at the beginning of the twentieth century. I have no idea whether such replicas have ever been produced, but they would be almost impossible to detect.

Brands

With the exception of Rolex and the recent boutique wristwatch makers, the major European and American companies were established in the nineteenth century or earlier. Consequently, collecting by brand name, including Rolex, encompasses both pocket watches and wristwatches.

For example, the Hamilton Watch Company of America manufactured railroad grade pocket watches, such as the model 940 in chapter 2, Figures 2-41 to 2-44, page 22, and navigation watches like the model 4992B in chapter 5, Figure 5-34, page 88. In the twentieth century the company changed production to include wristwatches and later earned additional fame for developing the first solid-state watch. The company history is found in Don Sauers's *Time for America—The Hamilton Watch Company 1892-1992*.

Similarly, Rotherhams in England initially manufactured fusee and going barrel pocket watches, like

Figure 6-24.

the 7-jewel fusee lever shown in Figure 6-24, and late in the nineteenth century it followed the American example and set up a machine-based production line. Most of its watches were ordinary, but it also made small numbers of very high-quality movements, before Swiss and American competition reduced the company to importing wristwatches like the one in chapter 4 as Figure 4-60, page 68.

Longines is a company that has successfully survived, manufacturing watches from 1865 until today. This company is somewhat underrated. Its founder Ernest Francillon and his coworker Jacques David were leaders in Swiss attempts to produce machine-made watches, and David was one of the driving forces behind the radical changes forced on the Swiss industry by their American competitors. Later, Longines was an important player in the development of wristwatches and providing chronographs for timing of sporting events.

Figure 6-25.

Figure 6-26.

Figures 6-27 and 6-28, far right.

Figure 6-29 and 6-30, far right.

The watch in Figures 6-25 to 6-28 is a Longines master navigation watch. This, together with similar watches by Hamilton and other companies, was mainly used to determine longitude in U.S. airplanes during World War II. (The problem of determining longitude, and its solution using very accurate clocks and watches, is described in many books. I think the best of these is Andrewes's *The Quest for Longitude*.) Figure 6-25 shows the watch suspended in its American-made protective case, which shields the movement from magnetism and jolts.

This navigation watch has a 24-hour dial that is marked "GCT" to show the watch is meant to display Greenwich Civil Time. The small hand and semicircular subdial below the 24-hour marker is an up-down dial that shows how many hours the watch has run since it was last wound. Such watches represent the peak of accuracy for pocket watches and very few wristwatches can better them. Because this watch was used by the U.S. Air Force, it is certainly relevant to American collectors.

Do not forget that every time you look at a watch you must evaluate every aspect of it. So you should have noted that, although the movement is clean and in excellent condition, one of the two case screws holding the movement to the case is missing (see arrow in Figure 6-28).

A more normal Longines watch is shown in Figures 6-29 and 6-30. This was used on a railroad, and the worn crown testifies to constant use over many years. It is in very good condition, however, because it was regularly serviced.

As I have noted, American movements were sold separately from their cases; consequently, they were highly decorated to make them more attractive to customers. One aspect of this was the jeweling war that took place when extra, unnecessary jewels were added to watches to entice potential buyers. The absence of jeweling wars in Switzerland means their watches of similar quality tend to have fewer jewels than their American counterparts, so we cannot assume the movement in Figure 6-30 is not as good as the preceding one, just because it only has 15 jewels. But, in fact, although it is a nice, very robust movement, the bimetallic balance is uncut and cannot be adjusted accurately.

Russian timepieces seem to be regarded as inferior and their prices are relatively low. However, some are extremely good and do not deserve what could be described as cultural aversion. Although not relevant here, Russian marine chronometers are as good as any other standard marine chronometer, but they currently sell for one third or less of the price of equivalent Western chronometers that are no better.

The Slava watch in Figures 6-31 and 6-32 is a good example of a sophisticated, high-quality wristwatch. This 21-jewel, manual wind movement has two main-springs (marked by the arrows) that enable weaker and more uniform springs to be used to improve accuracy.

Unfortunately I do not know of a good book on Russian watches; the books I have seen have endless photographs of different dials without much useful information.

Figure 6-31.

Figure 6-32.

Precision Watches

At the other extreme to dollar watches are chronometers and half chronometers. If you wish to collect these very accurate watches, your choices are somewhat limited. They were made in relatively small numbers and their cost, now as when new, is high.

Two examples of the acme of precision watches are shown in Figures 6-33 and 6-34. The pocket chronometer movement in Figure 6-33 is signed "Arnold Chas Frodsham" and was made by Charles Frodsham about 1845. It has a fusee, spring detent escapement (the end of the detent is at **a** and the escape wheel at **b**), and 13 jewels; both the 4th and escape wheels have end stones, and the 3rd wheel is not jeweled. The balance spring is a free sprung overcoil spring. The built-in, folding key (at **c**) is definitely a later addition, because it covers part of the inscription, and it was probably added when the movement was recased.

The half chronometer in Figure 6-34, made by Settle & Co. in 1939, is in its original 18-carat gold case. It is pendant wound with a going barrel and has a right angle lever escapement, free sprung overcoil balance spring, and 19 jewels.

Both would have been made and finished very precisely and then carefully adjusted over many weeks to obtain the best possible performance. Except for the flowery signatures and the engraved balance cock on the Frodsham watch, both

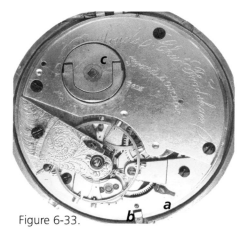

Figure 6-33.

Figure 6-34.

movements are quite plain. All the watchmaker's skill has been directed to quality rather than unnecessary decoration. The jeweling also is strictly functional with slow moving pivots running in brass holes. These free sprung pocket chronometers are superb watches and Thomas Earnshaw demonstrated that they can have a performance equaling that of marine chronometers.

The pivoted detent chronometer in Figure 6-35 was made by Ulysse Breting in Switzerland and dates from circa 1860. This has also been recased. It was originally key wound and set, but someone has added a keyless winding mechanism, although leaving it key set. It is a 19-jewel movement with cap jewels on the balance, escape wheel, and detent. It is very similar to the pivoted detent movement in Figure 4-37, page 63, except the Breting watch has a helical balance spring with its coils forming a cylinder.

Figure 6-35.

This conversion is a good example of the problem I raised when talking about reversible repairs. There are three serious difficulties for someone who wants to return this watch to its original form in an appropriate case. First, a new hole has been drilled through the barrel bridge for the screw **a**. Second, a different screw has been used at **b** to hold down both the bridge and the keyless mechanism; not only has the original screw been lost but this new screw may have a different thread. And third, the square on the barrel arbor at **c** may have been shortened so that a key cannot be used on it. Sadly, I suspect some damage has been done that cannot be undone, and it is unfortunate that this unsatisfactory and pointless modification has been made.

Despite the fact that a balance spring with a regulator cannot be adjusted as well as a free sprung balance spring, both of these pivoted detents (chapter 4, Figure 4-37, page 63, and Figure 6-35) have regulators. Consequently, I think they are inferior to the Frodsham and Settle watches.

There are two useful things to notice about these four watches. First, all of them have fewer jewels than American railroad watches; indeed, the best of them only has 13 jewels. Just how many jewels are necessary is debatable. I wonder if the Longines master navigation watch has 21 jewels because Longines wanted it that way or was forced to add additional jewels to satisfy the American contract for their purchase (Figures 6-27 and 6-28, page 111). Second, the micrometer regulator of the Ulysse Breting watch is hard over on the "fast" side (Figure 6-36). When I tested it with the regulator in the middle, it ran about seven minutes slow per day! Clearly, there is something very wrong with this watch, but more importantly there is a moral for collectors: If the regulator is not in the center, you should suspect something is definitely wrong with the movement.

At a somewhat lower level of accuracy are the two watches in Figures 6-37 and 6-38.

Figure 6-36.

Figure 6-37.

Figure 6-38.

Figure 6-37 shows a movement signed "Ebenezer Taylor London" and made about 1850. It has 21 jewels, but the number of jewels is only a part of the story, as we have seen. Just as American companies used jewels as aids to selling movements, the English were not above doing likewise, and the obvious example is the use of Liverpool windows (very large pivot jewels in chatons). This movement has pointless but beautiful jewels on the fusee arbor and no jewels on the center arbor, which is hidden under the balance cock and of no interest to a prospective buyer! However, this watch, like its dazzling American counterparts, has the signs of a fine piece capable of good accuracy.

Figure 6-38 shows a movement signed "George Fivey, Dublin." This was purchased as a bare movement off the Internet without seeing a photograph of the movement; in doing so I have broken my cardinal rule of never buying a watch unless I have seen at least one good picture of the movement. But I decided to take a gamble on it anyway, because there was a photograph of it with its dust cap on (like that in Figure 2-25 of chapter 2, page 18) showing that the shape of the balance cock was interesting; and there was a view under the dial showing that there were at least two jewels.

There are two things that stand out. First, the broad, flat, steel balance is typical of cheap English verge watches, like the one in chapter 3 as Figure 3-6, page 37. However, such watches are not jeweled and this one is. Second, the shape of the balance cock suggests it was made between 1800 and 1810; compare it with those shown in chapter 3, Figures 3-13 to 3-15, pages 40 and 41, and Figure 6-54, page 118. This is confirmed by Loomes's *Watchmakers and Clockmakers of the World,* which notes Fivey working between 1782 and 1812 when he died. So, except for a few jewels, this looks rather uninteresting. But when I had a chance to examine it, I found a very interesting watch.

First, it has a bimetallic regulator for temperature compensation, like that in chapter 5, Figure 5-31. But this one, seen clearly in Figure 6-39 is composed of two curved parts screwed together where they meet. Second, it is not a verge watch; it has a Savage two-pin lever escapement; you will find this escapement described in several books including Cutmore's *The Pocket Watch Handbook.* Figure 6-40 shows the two tiny pins on the roller (marked by the arrows) and the slot between them on the edge of the roller. The movement has 12 jewels, one less than an equivalent conventional lever escapement watch, with an impulse jewel on the roller. Third, the movement has the serial number 376, but the underside of the balance cock is stamped with the number 1028 (Figure 6-39). We can be fairly sure that 1028 is the serial number assigned by the actual movement maker, and George Fivey was just the retailer.

The last point is the most curious. According to Cutmore and other authors, Savage invented his version of the lever escapement around 1814-1815. But by that time, according to Loomes, Fivey was dead!

Figure 6-39.

Figure 6-40.

To add to my confusion, Figure 6-41 shows the escape wheel and the lever. At the end of the lever there is a large arc of steel (see arrow). This is a counterweight added so that the lever is balanced around its arbor. But it simply should not be present, because there is nothing at the other end of the lever that needs to be counterbalanced! However, the rack lever escapement has exactly this style of lever, because the other end is the same shape except that it has teeth to mesh with a pinion on the balance staff. So I was tempted to conclude that this watch originally had a rack lever escapement and some time after Fivey died it was changed to a Savage two-pin lever escapement. In doing so, the watchmaker kept the original lever, cutting and filing down the other end so that it had just two

Figure 6-41.

projections that fit around the pins on the balance staff. (Note that this movement does not have a slot on the top plate, as in the case of the rack lever shown in chapter 3 as Figures 3-13 and 3-18, pages 40-41. Not all rack lever watches have this slot, and the normal way to identify the escapement is by the counterweighted lever.)

But I was wrong because I had been careless. In Figure 6-41 it is quite clear that the lever from the arbor to the balance staff is very long and so it does need a counterweight to balance it. In which case there is no reason at all to think it has been converted. And the date? Well, reference books like that by Loomes do have mistakes, and, even if the dates are correct, the business may have continued after Fivey's death.

I originally included Fivey's watch in this section because of the bimetallic regulator for temperature adjustment; although not as effective as a compensation balance, watches with this feature have been designed for improved accuracy. However, it is actually more interesting for what you can learn by careful examination. What at first glance looked like a cheap verge watch turned out to have an early, very interesting, Savage lever movement, which in turn is really a modified rack lever, and, after more careful examination, is not a modification at all!

Figure 6-42.

Of course, collecting "accuracy" can be inverted and you might be more interested in ordinary watches. The two movements in Figures 6-42 and 6-43 are typical of ordinary English and American output. Both have lever escapements (right angle for the English and straight line for the American) and have not been adjusted. The American watch (which is the same watch as in chapter 5, Figure 5-26, page 83) could be adjusted, because it has a cut bimetallic balance and overcoil balance spring, but the English watch could not be adjusted. Both are solid, well-made watches produced in large numbers for ordinary people who did not need perfect timekeeping. The word "patent" on the English watch is potentially interesting, suggesting it may have some unusual feature, but sometimes the word was used simply as a sales aid.

Figure 6-43.

Figures 6-44 and 6-45 show the dial and movement of another ordinary English fusee watch, made in 1883. The only significant difference between this and the previous English watch is that this movement is oversprung; the regulator is placed over the balance. It also attractively announces that it was made by "Taylor

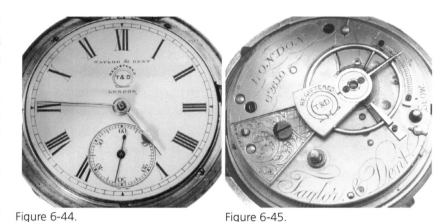

Figure 6-44.

Figure 6-45.

& Dent." Now, Edward John Dent was a very famous English maker, but the Dent in "Taylor & Dent" has nothing whatever to do with him. The name Dent has been used on many watches, including many crude Swiss fakes, to enhance the appeal of otherwise uninteresting watches. It seems it still works today because the prices achieved on the Internet for fake Dent watches are excessive. E. J. Dent also sold many ordinary watches that are no better than ordinary watches with other names on them.

Dent died in the 1850s and one part of the business was run by his wife Elizabeth under the name E. Dent. So a watch signed E. Dent has very little to do with the famous E. J. Dent.

Special Markets

Some watches have been manufactured for specific markets, and these watches were designed to satisfy the desires and needs of people in particular countries.

The most notable special market was Turkey. From the eighteenth century (and earlier) English and Swiss makers produced Turkish market watches with Arabic numeral dials. The early watches with verge escapements were often enclosed in three or four cases made of different materials. The later watches, like the one in Figures 6-46 and 6-47, were highly jeweled and had ornately engraved plates and cocks. This watch has 23 jewels, even more than the Longines half chronometer and most American railroad watches. Its dial is signed "K. Serkisoff & Co. Constantinople," but there is no inscription on the movement. The case is silver and has both Swiss and English hallmarks (which date the watch to 1895) and a trademark with the name Billodes.

The Roskopf watch described in chapter 4 (Figure 4-45, page 65) is also a Turkish market watch, as you can see from its dial in Figure 6-48, but it is clearly inferior.

Figure 6-46.

Figure 6-47.

Figure 6-48.

Before commenting on the Serkisoff watch, look at Figures 6-49 and 6-50. This watch has Russian Cyrillic text on the dial and case, but the case is also signed "Georges Favre-Jacot Locle." Again, it has 23 jewels, a straight-line lever escapement, an uncut compensation balance, and a flat balance spring. Indeed, if you ignore the decoration, both movements are almost identical, which actually is not surprising because both were made by the same company.

In 1865 Georges Favre-Bulle set up a watch factory in the Billodes district of the Swiss town Le Locle, which later became the famous Zenith watchmaking company. The trademark on the Serkisoff watch was registered in 1884, and the trademark on the Russian watch was registered in 1880.

Both of these watches have a straight-line lever escapement with an uncut compensation balance and a flat balance spring. The simple regulator and uncut balance mean that these watches cannot be and never were accurately adjusted; consequently, the jeweling is excessive. Although a basic 15 jewels would be justified, all the extra end stones are pointless additions. So despite having more jewels, neither of these watches is anywhere near as good as the precision watches in the previous section. Figures 6-51 and 6-52 is a typical Chinese market watch. These watches have sweep seconds hands and most have a Chinese duplex escapement. If you are careless,

Figure 6-49.

Figure 6-50.

you will not have noted that the cock under the balance is completely wrong, having no engraving. At some time the original Chinese duplex escapement was replaced by a cylinder escapement, requiring the watchmaker to fit a new cock. So this watch is a poor example of the type. These watches were designed and marketed to satisfy specific customer needs in Turkey, Russia, and China. In Turkey and China highly ornamented and jeweled watches were desired, although each type is completely different. In contrast, it would appear that Russian customers preferred a more austere but also over-jeweled movement. So how good are these watches? Pretty average is the answer. Despite being well made, such watches were more for show than good timekeeping and are very much inferior to the chronometers we have looked at.

Figure 6-51.

Figure 6-52.

Verge Watches

The verge escapement has been used in watches right from the very beginning, about 1480, to about 1850, a span of nearly 400 years. Verge watches made between 1750 and 1850 are quite common, and we have seen several English and continental examples already (see chapter 3, Figures 3-6, 3-8 to 3-11, and 3-27 [pages 37-40]; chapter 4, Figure 4-84 [page 72]; and chapter 5, Figure 5-12 [page 78]). As I have noted, the hand-engraved and -pierced balance cock usually indicates a verge watch before 1800. However, this type of cock was used on all watches, and you may come across ones with the cylinder escapement or a variant of the verge like the Ormskirk chaffcutter.

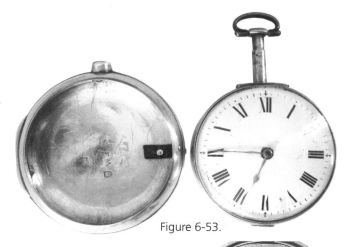

Figure 6-53.

Figure 6-54.

Earlier I mentioned that the shape and piercing of the cock is a good yet rough guide to age. The watches in Figures 6-53 to 6-56 are both pair case watches. The first, which is in a case dated 1808, has a cock that stylistically fits neatly between the two cocks in chapter 3 as Figures 3-13 and 3-14, pages 40-41. Note that the movement has no jewels. Instead, the balance cock has a blind hole that does not go completely through the cock.

If you handled this watch, it may give the impression that it is a marriage; although the maker, Charles O'Reilly of Dublin, worked before 1820 and the case has the "correct" date, the movement catch does not lock into the edge of the case. However, closer inspection reveals the edge of the case is badly worn and the case and movement definitely belong together.

The watch in Figures 6-55 and 6-56 should have an earlier date than the one we have just looked at, having a typical English pierced cock. But the case is dated 1830. Both the maker, William Venables, and the casemaker, Benjamin Norton, worked circa 1830. Consequently, if you look casually at a watch, you may deduce a significantly incorrect date for it; every watch must be carefully examined and all its elements considered.

In addition to noting the shape of the balance cock, the pendant style (long and thin, short and fat, and so on) is also a useful guide.

Figure 6-56.

Figure 6-55.

All verge watches look pretty much the same. The movement in Figure 6-57 is like any other continental verge and it has at least one less jewel than the watch in chapter 4 as Figure 4-84, page 72. However, hidden between the plates is an example of high-quality workmanship. Both the vertical contrate wheel arbor and the horizontal escape wheel arbor have waists in them so that they can be placed closer together without interfering with each other. Part of the top pivot of the contrate wheel arbor also is visible under the top plate and it is exquisitely small (Figure 6-58).

coqueret

Figure 6-57.

Figure 6-58.

There are actually two other indications that this might be an above average watch. The maker is Charles Le Roy, a French maker of high repute, and the case is two-color gold with a translucent blue enamel back, within which silver and gold decorations (paillonnes) are embedded (Figure 6-59).

However, cases and general appearance can be misleading and many quite ordinary verge watches will look very elegant. So it is still necessary to check the fine detail of such watches, as you should do with any watch.

Social Status

On the whole, collecting and wearing watches do not go together. But I suspect everyone likes to wear a nice watch and I sometimes use one of the pieces in my collection. The choice is simple: the watch that people who know nothing about horology will find stunning! I have three watches that are suitable.

The watch shown in Figure 6-60 is very pretty. The unblemished hunter case is engine turned 18-carat gold and reveals a very delicately painted enamel dial. The movement (Figure 6-61) is beautiful in its simplicity, with a little damaskeening around the edge and 16 ruby jewels. A cursory glance, which is all the layperson is usually capable of, shows it to be elegant and delightful.

Of course, the collector looks a little deeper. Although this is a high-quality Swiss bar movement, a careful examination reveals it has a bimetallic balance, but the balance rim is deceptively cut half way through to give the impression that it could be adjusted for temperature. So more than a quick look is needed; this no-name movement is good, but not exciting.

Figure 6-59.

Figure 6-60.

Figure 6-61.

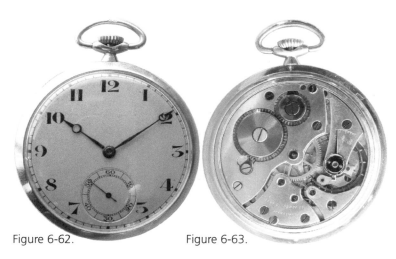

Figure 6-62.   Figure 6-63.

So I have changed my mind and decided to wear the better quality Marvin watch in Figures 6-62 and 6-63. Although the case is only 9-carat gold (which I won't tell anyone) the watch is very attractive and has a 17-jewel adjusted movement. Being knowledgeable, you will be aware that something is not quite right because of the position of the regulator.

The watch in Figures 6-64 and 6-65 has an 18-carat gold pair case, which is dated to 1820 by the hallmarks. The movement is a Massey lever with 13 jewels, an uncut polished-steel balance, and a stop lever. Although not the best Massey lever movement in my collection, it is still a fine example of an early lever watch. More importantly, in a social context, it is guaranteed to elicit gasps of admiration, mainly because of the very large amount of gold in the two cases.

Figure 6-64.   Figure 6-65.

Social status watches can be either wristwatches or pocket watches. However, unless you take some action, people around you will ignore both. Because simply thrusting a watch in someone's face is a bit rude, tactfully introducing the subject of horology into the conversation is necessary, after which you can flourish your timepiece and elicit the desired response. The best watch for this purpose is the one with the most highly decorated and easily accessible movement, so in the future my favorite will be my Hamilton 940 or Longines navigation watch.

Functions

Collecting by function allows you to concentrate on some particular aspect of the watchmaker's craft; chronographs, repeaters, navigation watches, and automatic mechanisms are a few examples. One interesting category is the center seconds stopwatch. These pocket watches are like the so-called doctor's watches where the stop lever acts on the escapement and halts the entire mechanism. Furthermore, there is no way to set or reset the seconds hand. Consequently, whenever they are stopped and started, they lose track of time, and so they are pretty well useless for timing anything! Also, the sweep seconds hand rotates once every minute, but the dial is marked in 1/5 seconds from 1 to 300! But despite these disad-

Figure 6-66.

Figure 6-67.

Figure 6-68.

Figure 6-69.

vantages, many of these large, heavy watches were produced in England and Switzerland at the end of the nineteenth century.

Figures 6-66 and 6-67 show the dials of two center seconds stopwatches. Both are in silver cases and they were made in 1896 and 1895, respectively.

The movements of these two watches, Figures 6-68 and 6-69, are almost identical. The one on the left is signed "Made for G. Cohen, Manchester." It has 15 jewels and uses a right-angle English lever escapement with pointed escape wheel teeth. This watch has a fusee; the barrel and fusee arbors are marked by arrows.

The movement in Figure 6-69 uses a Swiss right-angle lever escapement and has 13 jewels; the lever arbor under the balance is not jeweled. Being pendant wound, the two arbors are for the barrel and the keyless mechanism joining the crown to it.

The watch signed Cohen was probably made in England, whereas the second watch is almost certainly Swiss.

A number of variations will be found. There are Swiss lever escapement, key-wound watches with an extra pinion acting as a dummy fusee to reverse the direction of winding. These were presumably manufactured for the English market. Some watches have cylinder escapements. Although I do not know the origin of these strange and rather pointless watches, all the ones I have appear to be for the English market, so they may have been manufactured to satisfy some peculiar whim of the English. To enhance their appeal, they quite often have misleading information on the dial; one watch dial announces "Marine Chronograph," implying sophistication and very high accuracy, neither of which is true!

Small Wristwatches

Wristwatches for women is another neglected area, perhaps indicating that most collectors are men. However, the term "small" is better because some watches made for men are as small as those made for women. As we have seen, early wristwatches also were often derived from pocket watches for women. Indeed, one of the following watches was worn by a man and all but one could have been.

The largest, Figures 6-70 and 6-71, is 30 mm (1-1/4 inches) in diameter and is a 17-jewel Longines watch made circa 1955 (dated by the serial number). Note that it has a shock-resistant jewel on the balance cock.

Figure 6-70.

Figure 6-71.

Figure 6-72.   Figure 6-73.

The Maximus watch in Figures 6-72 and 5-73 is 28 mm in diameter and has a 15-jewel, adjusted movement. The jewels, which are pressed into the cocks, are surrounded by square, gold-colored sinks, which is rather pointless decoration.

Figure 6-74.   Figure 6-75.

The watch in Figures 6-74 and 6-75 has a 27 mm x 18 mm case holding a 15-jewel adjusted movement. The dates of this watch and the Maximus are unknown, but they are probably circa 1940.

The Ultima watch in Figures 6-76 and 6-77 is 25 mm in diameter.

Figures 6-76 and 6-77, far left.

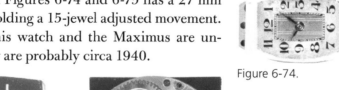

The last watch (Figures 6-78 and 6-79) is tiny. This automatic Omega has a movement about 13 mm or 1/2 inch in diameter. The serial number dates it to circa 1950. Note that the shock-resistant jewel on the balance cock is missing. This and the Atomic watch in Figures 6-74 and 6-75 are the only ones that a man would unlikely wear because of their size and style.

Figure 6-78.   Figure 6-79.

All of these watches have uncut monometal balances with screws. It is not possible to determine, just by looking at them, whether these balances and their balance springs are made of the special metals necessary to adjust them for temperature.

However, while a wristwatch is being worn, it is kept at a reasonably stable temperature by body heat, and temperature adjustment may have been considered relatively unimportant. All also have painted metal dials and straight-line Swiss lever escapements.

If we compare wristwatches of reasonable quality, then two points become clear. First, large numbers, probably thousands of signatures, have been used. But most of these watches will have originated from one of the ébauche factories and will have a maker's trademark under the dial or elsewhere. Second, manufacturing methods and watch design have reached a high degree of standardization so that the movements are fundamentally similar. There are differences in detail and quality, but not much variation in style.

Now You Are Not a Novice

My job is nearly done. And it wasn't difficult, because I have enjoyed writing this book and rediscovering the watches in my collection, which I have pulled out of drawers to photograph. The only thing left is to summarize what I have tried to do.

My aim has been simple: to explain one way in which you can progress from being a novice to being a *competent* watch collector; and I think it is the best way, although I am sure others will disagree. I have assumed you began with an interest in and a feeling for watches. Then I have suggested how you can nurture that interest by progressively expanding your practical experience and knowledge, starting with simple things like opening cases and ending with the ability to assess and repair watches, in theory if not in practice.

Although it has probably taken only a little time to read this book, your development as a collector, and mine, will take much longer, if only because it will take you more than a few days to open and examine a hundred watches and read 15 or so books, and it will take me some time to read another 15 books and look at yet another hundred watches. It is a never ending "pilgrim's progress" that is a labor of love and pleasure, filled with moments of joy when you hold a beautiful watch or read a delightful book.

When you decide you are competent is up to you, but one day you will feel comfortable with your progress and realize you have knowledge and ability that you can and want to share with others. You will be aware of a gradual shift from being a listening learner to a talking and competent collector.

I stress the word competent because I do not think the journey I have mapped out is sufficient to reach the point where you might consider yourself an *expert*. But then, I have no idea what an expert is! All I know is that there is no quick, let alone instant way to reach that goal, and after more than 20 years I think I still have some way to go. But maybe you will get there before I do. However, one thing that you will find is that the pleasure you gain from studying time will make time itself fly!

Appendix A—A Beginner's Library

Unlike many watch book lists, which seem to be almost random selections from the vast literature on horology, the following lists contain books that I believe you should *own* and *read*.

I have split the books into two lists. The first list contains those to be read to develop knowledge and understanding. I have organized them roughly in the order I think you should buy and read them. I have listed my bibliography *Mechanical Watches* first, not because of an inflated self-opinion but because I have stressed throughout the need to read and study. But if you want to read books other than those listed here, how do you find out what is available? I am suggesting you read *Mechanical Watches* simply because it will help you find answers to that question, and it is free. Although it is an alphabetical list of more than 2,800 books, it is worth skimming through from front to back, stopping to read about books that seem interesting and that you may want to buy to further your knowledge.

The order is rather arbitrary. As I have said in the last chapter, the process of education necessary to become a competent collector takes much longer than the time necessary to read this guide. So I would expect that reading these other books will go on in parallel with the development of your practical experience. Indeed it is essential. There is no strict order and no need to finish one before embarking on the next. One of the joys of collecting is that you are your own master, and you can develop in whatever way complements and nurtures your pleasure and interests.

Because I very strongly believe collectors should have some understanding of watch repair (and hopefully watchmaking as well) I do not regard the two repair books by Fletcher and Whiten as optional.

The second list contains books to be referenced and these books are organized alphabetically. Which of these books you buy depends on the area of horology in which you decide to specialize. Because there are not many of them and most are easy to obtain, you may decide to get all or most of them. Even if you focus on American watches, for example, you will meet enough English and Swiss watches to find some need for books covering those countries.

Books to Read

The numbers in parentheses are the pages where the book is mentioned.

Watkins, Richard. *Mechanical Watches: An Annotated Bibliography of Publications since 1800*. Australia: the author, 2011. www.watkinsr.id.au (*3, 6, 26, 92, 103*)

Cutmore, Max. *The Pocket Watch Handbook*. London: Bracken Books, 1985. (*31, 57, 114*)

Cutmore, Max. *Watches 1830-1980*. Newton Abbot: David & Charles, 2002. (*31, 42, 44, 53, 104, 108*)

Britten, F. J. *Watch and Clockmakers' Handbook, Dictionary and Guide*. London: E. & F. N. Spon, 1987. (*3, 53, 57, 100*)

Landes, David. *Revolution in Time*. Cambridge, MA: Harvard University Press, 2000. (*105*)

Meis, Reinhard. *Pocket Watches from the Pendant Watch to the Tourbillon*. Atglen, PA: Schiffer Publishing, 1999. (*31, 42, 44, 57, 63, 64, 74, 106*)

Kahlert, Muhe, and Brunner: *Wristwatches: History of a Century's Development*. Atglen, PA: Schiffer Publishing, 1999. (*31, 44, 45, 53, 57, 62, 74*)

Baillie, Ilbert and Clutton. *Britten's Old Clocks and Watches and Their Makers*, 9th Ed. London: Bloomsbury Books, 1986. (*42*)

Shenton, Alan. *Pocket Watches 19th & 20th Century*. Suffolk: Antique Collectors Club, 1996. (*42, 44*)

Hoke, D. R. *The Time Museum Historical Catalogue of American Pocket Watches*. Rockford, IL: The Time Museum, 1991. (*31, 43, 104*)

Harrold, Michael. *American Watchmaking: A Technical History of the American Watch Industry 1850-1930*. Columbia, PA: National Association of Watch and Clock Collectors, 1984. (*31, 43, 104*)

Fletcher, D. W. *Watch Repairing as a Hobby*. Johnson City, TN: Arlington Book Co., 1986. (*92*)

Whiten, Anthony. *Repairing Old Clocks and Watches*. London: NAG Press, 1996. (*49, 92*)

Baillie, G. H. *Watches—Their History, Decoration and Mechanism*. London: NAG Press, 1979. (*42*)

Lecoultre, Francois. *Guide to Complicated Watches*. Neuchatel: A. Simonin, 1993. (*74*)

Reference Books

Banister, Judith. *English Silver Hall-Marks*. Slough, UK: W. Foulsham & Co., 1990. (*10*)

Buffat, Eugene. *History and Design of the Roskopf Watch*. Kingston Beach, Australia: Richard Watkins, 2007. (Available from www.watkinsr.id.au) (*66*)

Chamberlain, Paul. *It's About Time*. London: The Holland Press, 1978. (*8, 58, 61, 104*)

Crossman, Charles. *A Complete History of Watch and Clock Making in America*. Donald Dawes, 2002. (*77*) (Or the first edition *The Complete History of Watchmaking in America*.)

Cutmore, Max. *Collecting and Repairing Watches*. Newton Abbot, Devon, UK: David & Charles, 1999. (*106*)

Ehrhardt, Roy and Meggers, Bill. *American Watches Beginning to End: Identification & Price Guide*, 1987, Kansas City, MO: Heart of America Press, 1987 (reprinted in 1998 and 2009). (*33, 42, 43*)

Loomes, Brian. *Watchmakers and Clockmakers of the World, Complete 21st Century Edition*. London: Robert Hale, 2006. (*34, 114*)

Priestley, Philip. *Watch Case Makers of England 1720-1920*. Columbia, PA: National Association of Watch and Clock Collectors, 1994. (*37*)

Pritchard, Kathleen. *Swiss Timepiece Makers 1775-1975*. Columbia, PA: National Association of Watch and Clock Collectors, 1997. (*34, 47, 71, 80*)

Tardy. *Dictionnaire des Horlogers Francais*. Paris: Tardy, 1971. (*34*)

Appendix B—Further Reading

In addition to my suggestions for a beginner's library, the following books have been mentioned in the text. Most of these books are either supplementary, covering similar material to my recommendations, or specialist, covering specific areas of watch making and collecting.

The fact that I have mentioned them means that they are relevant to some aspect of collecting that I have discussed, and they are in no way comprehensive. For example, only two makers, Hamilton and Rolex, are included because I have referred to them. There are, of course, many other books that cover watches from particular makers, and my choice of these two in no way implies any preference. Naturally you will acquire books appropriate to your own interests.

The numbers in parentheses are the pages where the book is mentioned.

Andrewes, William. *The Quest for Longitude*. Cambridge, MA: Harvard University, 1996. (*101, 111*)

[Anon]. *Official Catalogue of Swiss Watch Repair Parts*, edition M. Switzerland: Les Fabricants Suisses D'horlogerie, 1949. (*38*)

Antiquarian Horological Society. *Your Time*. England: Antiquarian Horological Society, 2008. (104)

Berthoud and Auch. *How to Make a Verge Watch*. Australia: Richard Watkins, 2005. (*101, 104*)

Brown, R. *Replica Watch Report*. USA: the author, 2004. (*78, 104*)

Bruton, Eric. *Collector's Dictionary of Clocks and Watches*. London: Robert Hale, 1999. (*3*)

Bruton, Eric. *Dictionary of Clocks and Watches*. New York: Bonanza Books, 1963. (This is the first edition of *Collector's Dictionary of Clocks and Watches*.) (*3*)

Bulova School of Watchmaking. *Training Units*. New York: Bulova School of Watchmaking. (*95*)

Carrera, R. *Hours of Love*. Lausanne: Scriptar, 1977. (*74*)

Cutmore, Max. *Pin Lever Watches*. England: D. H. Bacon, 1991. (*108*)

de Carle, Donald. *Watch and Clock Encyclopedia*. London: NAG Press, 1995. (*3*)

de Carle, Donald. *With the Watchmaker at the Bench*. London: Isaac Pitman & Sons, 1946. (*49*)

Daniels, George. *Watchmaking*. London: Philip Wilson, 2002. (*105*)

David, Jacques. *American and Swiss Watchmaking in 1876*. Australia: Richard Watkins, 2002. (*104*)

Dowling and Hess. *The Best of Times—Rolex Wristwatches: An Unauthorized History*. Atglen, PA: Schiffer, 2001. (*27, 104*)

Edidin, Michael. "English Watches for the American Market." *NAWCC Bulletin*, No. 280 (October 1992): pp. 515-544 and No. 281 (December 1992): pp. 659-693. (*79*)

Fried, Henry. *The Watch Repairer's Manual*. Harrison, OH: American Watchmakers Institute, 1999. (*95*)

Fried, Henry. *Bench Practices for Watch and Clockmakers*. New York: Columbia Communications, 1994. (*95*)

Jaquet, E. and Chapuis, A. *The History of the Self-Winding Watch 1770-1931*. London: B. T. Batsford, 1956. (*53*)

Jendritzki, H. *The Swiss Watch Repairer's Manual*. Switzerland: Scriptar, 1977. (*95*)

Kochmann, Karl. *Clock and Watch Trademark Index—European Origin*. West Sacramento, CA: Clockwork Press, 2001. (*47*)

Lucchina and Perkins. *The Watchmakers' Staking Tool*. Lebanon, NH: Witman Press, 1987. (*96*)

Perkins, A. *The Watchmakers Lathe and How To Use It*. Harrison, OH: American Watchmakers-Clockmakers Institute, 2003. (*100*)

Sauers, D. *Time for America—The Hamilton Watch Company 1892-1992*. Lititz, PA: Sutter House, 1992. (*110*)

Schaeder, Albin. *The Proletarian Watch*. Sweden: Urmakaren Publishing Company, 2006. (*108*)

Shugart, Engel, and Gilbert. *Complete Price Guide to Watches*. Cleveland, TN Cooksey Shugart Publications. (*33, 37, 61*)

Vigniaux. *Practical Watchmaking*. Kingston, Tasmania: Richard Watkins, 2011. (Available from www.watkinsr.id.au) (*105*)

Watkins, Richard. *The Repeater*. Kingston, Tasmania: the author, 2011. (Available from www.watkinsr.id.au) (*74*)

Weiss, Leonard. *Watch-Making in England 1760-1820*. London: Robert Hale, 1982. (*105*)

Index

To Experience Adventures in TIME...

It All Starts with Membership

The National Association of Watch and Clock Collectors, Inc. (NAWCC) is an international non-profit association serving more than 17,000 members and 150 chapters and dedicated to preserving and stimulating interest in horology, the art and science of time. Our members are enthusiasts, students, educators, casual collectors, businesses, and professionals, who love learning about the clocks and watches they preserve, study, and collect. Members share their interests with other members and establish friendships around the world.

Membership Advantages

• Stay informed with the bi-monthly *Watch & Clock Bulletin,* an educational journal, and the *Mart & Highlights,* a buy/sell/news publication.

• Go online for research tools and videos: NAWCC.org features all *Watch & Clock Bulletin* content back to 1943, NAWCC books and instructional videos, and much more for members.

• Buy, sell, and learn at regional buying and selling venues, and attend programs on all aspects of horology.

• Keep in touch with our bi-monthly electronic newsletter—*eHappenings*

• Meet terrific people at local and special interest chapters

• Visit for free the National Watch & Clock Museum in Columbia, PA.

• Use your membership for free or discounted admission to over 250 museums and science centers.

Become a member today and begin your exploration of the fascinating world of horology.

Apply online at www.nawcc.org

Join the NAWCC

Become a member today!
Mail this application, apply online at www.nawcc.org, or call 1-877-255-1849 or 1-717-684-8261.

*Required fields

_____
*Print Name

_____
Company Name (optional)

_____
*Street

_____
*City

_____
*State/Province/Country   *Zip/Postal Code

(   )   (   )
Ph.: Home   Work

(   )   (   )
Cell   Fax

_____
Email

❑ I agree to abide by the NAWCC Member Code of Ethical Conduct (see nawcc.org to review).

*Are you a former member of NAWCC? ❑ Yes   ❑ No

_____
If "yes," your membership no.

     /   /
Date of Birth Confidential for verification purposes. Required for Youth Membership.

_____
Occupation

_____
How did you learn about the NAWCC?

Interest: ❑ Clocks ❑ Wristwatches ❑ Pocket Watches ❑ Museum

From:
If this is a gift membership, print your name above and a gift card will be included with the membership card mailing.

Send this application with payment to:
NAWCC, Inc., 514 Poplar Street, Columbia, PA 17512-2130
Annual dues:
❑ **Individual $70** (mailed pubs.)   ❑ **Individual $60** (electronic publications)
❑ **Business $125** (mailed pubs.)   ❑ **Assoc./Youth $20** (electronic pubs.)
                                                          (Spouse) / (Under 18)
                                              ❑ **Student $35** (electronic pubs.)
                                                 Proof of enrollment required for student membership. Please call for information.

Payment:
❑ **Check enclosed (U.S. bank only)** ❑ **Intl. Money Order**
❑ **Visa** ❑ **MasterCard** ❑ **Discover** ❑ **American Express**

_____
Credit Card No.

     /
Exp. Date   Security Code

_____
Cardholder's Name   Amt. to be charged

_____
Signature

CPSIA information can be obtained
at www.ICGtesting.com
Printed in the USA
LVHW072015190620
658541LV00012B/283